T0291682

CAMBRIDGE LIBRARY COLLECTION

Books of enduring scholarly value

Life Sciences

Until the nineteenth century, the various subjects now known as the life sciences were regarded either as arcane studies which had little impact on ordinary daily life, or as a genteel hobby for the leisured classes. The increasing academic rigour and systematisation brought to the study of botany, zoology and other disciplines, and their adoption in university curricula, are reflected in the books reissued in this series.

Mendel's Principles of Heredity

William Bateson (1861-1926) began his academic career working on variation in animals in the light of evolutionary theory. He was inspired by the rediscovery in 1900 of the 1860s work on plant hybridisation by the Austrian monk Gregor Mendel to pursue further experimental work in what he named 'genetics'. He realised that Mendel's results could help to solve difficult biological questions and controversies which others had glossed over, and to challenge the status quo in evolutionary studies. Annoyed by the 'apathetic' stance of his evolutionist colleagues, and incensed by the Oxford professor Raphael Weldon's scathing critique of Mendel in a journal article, Bateson incorporated an English translation of Mendel's work into this 1902 book along with a thorough defence of Mendel's statistical experiments and the principles of heredity derived from them. His book is an impassioned appeal for scientists to adopt this 'brilliant method' which he felt could revolutionise both scholarship and industry. This volume also contains, in an appendix, the original German texts of Mendel's Versuche über Pflanzenhybriden, as published in Leipzig in 1901.

Cambridge University Press has long been a pioneer in the reissuing of out-of-print titles from its own backlist, producing digital reprints of books that are still sought after by scholars and students but could not be reprinted economically using traditional technology. The Cambridge Library Collection extends this activity to a wider range of books which are still of importance to researchers and professionals, either for the source material they contain, or as landmarks in the history of their academic discipline.

Drawing from the world-renowned collections in the Cambridge University Library, and guided by the advice of experts in each subject area, Cambridge University Press is using state-of-the-art scanning machines in its own Printing House to capture the content of each book selected for inclusion. The files are processed to give a consistently clear, crisp image, and the books finished to the high quality standard for which the Press is recognised around the world. The latest print-on-demand technology ensures that the books will remain available indefinitely, and that orders for single or multiple copies can quickly be supplied.

The Cambridge Library Collection will bring back to life books of enduring scholarly value across a wide range of disciplines in the humanities and social sciences and in science and technology.

Mendel's Principles of Heredity

A Defence, with a Translation of Mendel's Original Papers on Hybridisation

William Bateson
Gregor Mendel

CAMBRIDGE UNIVERSITY PRESS

Cambridge New York Melbourne Madrid Cape Town Singapore São Paolo Delhi

Published in the United States of America by Cambridge University Press, New York

www.cambridge.org
Information on this title: www.cambridge.org/9781108006132

This edition first published 1902
This digitally printed version 2009

ISBN 978-1-108-00613-2

MENDEL'S
PRINCIPLES OF HEREDITY

London: C. J. CLAY AND SONS,
CAMBRIDGE UNIVERSITY PRESS WAREHOUSE,
AVE MARIA LANE,
AND
H. K. LEWIS, 136, GOWER STREET, W.C.

Glasgow: 50, WELLINGTON STREET.
Leipzig: F. A. BROCKHAUS.
New York: THE MACMILLAN COMPANY.
Bombay and Calcutta: MACMILLAN AND CO., LTD.

GREGOR MENDEL

Abbot of Brünn

Born 1822. Died 1884.

From a photograph kindly supplied by the Very Rev. Dr Janeischek, the present Abbot.

MENDEL'S
PRINCIPLES OF HEREDITY

A DEFENCE

BY

W. BATESON, M.A., F.R.S.

WITH A TRANSLATION OF MENDEL'S ORIGINAL PAPERS ON HYBRIDISATION.

CAMBRIDGE:
AT THE UNIVERSITY PRESS.
1902

Cambridge:
PRINTED BY J. AND C. F. CLAY,
AT THE UNIVERSITY PRESS.

PREFACE.

IN the Study of Evolution progress had well-nigh stopped. The more vigorous, perhaps also the more prudent, had left this field of science to labour in others where the harvest is less precarious or the yield more immediate. Of those who remained some still struggled to push towards truth through the jungle of phenomena: most were content supinely to rest on the great clearing Darwin made long since.

Such was our state when two years ago it was suddenly discovered that an unknown man, Gregor Johann Mendel, had, alone, and unheeded, broken off from the rest—in the moment that Darwin was at work—and cut a way through.

This is no mere metaphor, it is simple fact. Each of us who now looks at his own patch of work sees Mendel's clue running through it: whither that clue will lead, we dare not yet surmise.

It was a moment of rejoicing, and they who had heard the news hastened to spread them and take the instant way. In this work I am proud to have borne my little part.

But every gospel must be preached to all alike. It will be heard by the Scribes, by the Pharisees, by Demetrius the Silversmith, and the rest. Not lightly do men let their occupation go; small, then, would be our wonder, did we find the established prophet unconvinced. Yet, is it from misgiving that Mendel had the truth, or merely from indifference, that no naturalist of repute, save Professor Weldon, has risen against him?

In the world of knowledge we are accustomed to look for some strenuous effort to understand a new truth even in those who are indisposed to believe. It was therefore with a regret approaching to indignation that I read Professor Weldon's criticism*. Were such a piece from the hand of a junior it might safely be neglected; but coming from Professor Weldon there was the danger—almost the certainty—that the small band of younger men who are thinking of research in this field would take it they had learnt the gist of Mendel, would imagine his teaching exposed by Professor Weldon, and look elsewhere for lines of work.

In evolutionary studies we have no Areopagus. With us it is not—as happily it is with Chemistry,

* *Biometrika*, I., 1902, Pt. II.

Physics, Physiology, Pathology, and other well-followed sciences—that an open court is always sitting, composed of men themselves workers, keenly interested in every new thing, skilled and well versed in the facts. Where this is the case, doctrine is soon tried and the false trodden down. But in our sparse and apathetic community error mostly grows unheeded, choking truth. That fate must not befall Mendel now.

It seemed imperative that Mendel's own work should be immediately put into the hands of all who will read it, and I therefore sought and obtained the kind permission of the Royal Horticultural Society to reprint and modify the translation they had already caused to be made and published in their Journal. To this I add a translation of Mendel's minor paper of later date. As introduction to the subject, the same Society has authorized me to reprint with alterations a lecture on heredity delivered before them in 1900. For these privileges my warm thanks are due. The introduction thus supplied, composed originally for an audience not strictly scientific, is far too slight for the present purpose. A few pages are added, but I have no time to make it what it should be, and I must wait for another chance of treating the whole subject on a more extended scale. It will perhaps serve to give the beginner the slight assistance which will prepare him to get the most from Mendel's own memoir.

The next step was at once to defend Mendel from Professor Weldon. That could only be done by following this critic from statement to statement in detail, pointing out exactly where he has gone wrong, what he has misunderstood, what omitted, what introduced in error. With such matters it is easy to deal, and they would be as nothing could we find in his treatment some word of allusion to the future ; some hint to the ignorant that this is a very big thing ; some suggestion of what it all *may* mean if it *be* true.

Both to expose each error and to supply effectively what is wanting, within the limits of a brief article, written with the running pen, is difficult. For simplicity I have kept almost clear of reference to facts not directly connected with the text, and have foregone recital of the now long list of cases, both of plants and animals, where the Mendelian principles have already been perceived. These subjects are dealt with in a joint Report to the Evolution Committee of the Royal Society, made by Miss E. R. Saunders and myself, now in the Press. To Miss Saunders who has been associated with me in this work for several years I wish to express my great indebtedness. Much of the present article has indeed been written in consultation with her. The reader who seeks fuller statement of facts and conceptions is referred to the writings of other naturalists who have studied the phenomena at first hand (of which a bibliography is appended) and to our own Report.

I take this opportunity of acknowledging the unique facilities generously granted me, as representative of the Evolution Committee, by Messrs Sutton and Sons of Reading, to watch some of the many experiments they have in progress, to inspect their admirable records, and to utilise these facts for the advancement of the science of heredity. My studies at Reading have been for the most part confined to plants other than those immediately the subject of this discussion, but some time ago I availed myself of a kind permission to examine their stock of peas, thus obtaining information which, with other facts since supplied, has greatly assisted me in treating this subject.

I venture to express the conviction, that if the facts now before us are carefully studied, it will become evident that the experimental study of heredity, pursued on the lines Mendel has made possible, is second to no branch of science in the certainty and magnitude of the results it offers. This study has one advantage which no other line of scientific inquiry possesses, in that the special training necessary for such work is easily learnt in the practice of it, and can be learnt in no other way. All that is needed is the faithful resolve to scamp nothing.

If a tenth part of the labour and cost now devoted by leisured persons, in this country alone, to the collection and maintenance of species of animals and plants which have been collected a hundred times before, were applied to statistical experiments in heredity, the result in a few years would make a revolution not only in the industrial art of the breeder but in our views of heredity, species and variation. We have at last a brilliant method, and a solid basis from which to attack these problems, offering an opportunity to the pioneer such as occurs but seldom even in the history of modern science.

We have been told of late, more than once, that Biology must become an *exact* science. The same is my own fervent hope. But exactness is not always attainable by numerical precision: there have been students of Nature, untrained in statistical nicety, whose instinct for truth yet saved them from perverse inference, from slovenly argument, and from misuse of authorities, reiterated and grotesque.

The study of variation and heredity, in our ignorance of the causation of those phenomena, *must* be built of statistical data, as Mendel knew long ago; but, as he also perceived, the ground must be prepared by specific experiment. The phenomena of heredity and variation are specific, and give loose and deceptive answers to any but specific questions. That is where our *exact* science will begin. Otherwise we may one day see those huge foundations of "biometry" in ruins.

But Professor Weldon, by coincidence a vehement preacher of precision, in his haste to annul this first positive achievement of the precise method, dispenses for the moment even with those unpretending forms of precision which conventional naturalists have usefully practised. His essay is a strange symptom of our present state. The facts of variation and heredity are known to so few that anything passes for evidence; and if only a statement, or especially a conclusion, be negative, neither surprise nor suspicion are aroused. An author dealing in this fashion with subjects commonly studied, of which the literature is familiar and frequently verified, would meet with scant respect. The reader who has the patience to examine Professor Weldon's array of objections will find that almost all are dispelled by no more elaborate process than a reference to the original records.

With sorrow I find such an article sent out to the world by a Journal bearing, in any association, the revered name of Francis Galton, or under the high sponsorship of Karl Pearson. I yield to no one in admiration of the genius of these men. Never can we sufficiently regret that those great intellects were not trained in the profession of the naturalist.

Mr Galton suggested that the new scientific firm should have a mathematician and a biologist as partners, and—soundest advice—a logician retained as consultant*. Biologist surely must one partner be, but it will never do to have him sleeping. In many well-regulated occupations there are persons known as "knockers-up," whose thankless task it is to rouse others from their slumber, and tell them work-time is come round again. That part I am venturing to play this morning, and if I have knocked a trifle loud, it is because there is need.

March, 1902.

* *Biometrika*, I. Pt. I. p. 5.

CONTENTS.

INTRODUCTION.

A DEFENCE OF MENDEL'S PRINCIPLES OF HEREDITY, 104—208.

ERRATA.

p. 22, par. 3, line 2, for "falls" read "fall."
p. 63, line 12, for "*AabbC*" read "*AaBbc*."
p. 66, in heading, for "OF HYBRIDS" read "OF THE HYBRIDS."

Note to p. 125. None of the yellow seeds produced by *Laxton's Alpha* germinated, though almost all the green seeds sown gave healthy plants. The same was found in the case of *Express*, another variety which bore some yellow seeds. In the case of *Blue Peter*, on the contrary, the yellow seeds have grown as well as the green ones. Few however were *wholly* yellow. Of nine yellow seeds produced by crossing green varieties together (p. 131), six did not germinate, and three which did gave weak and very backward plants. Taken together, this evidence makes it scarcely doubtful that the yellow colour in these cases was pathological, and almost certainly due to exposure after ripening.

THE PROBLEMS OF HEREDITY AND THEIR SOLUTION*.

An exact determination of the laws of heredity will probably work more change in man's outlook on the world, and in his power over nature, than any other advance in natural knowledge that can be clearly foreseen.

There is no doubt whatever that these laws can be determined. In comparison with the labour that has been needed for other great discoveries we may even expect that the necessary effort will be small. It is rather remarkable that while in other branches of physiology such great progress has of late been made, our knowledge of the phenomena of heredity has increased but little; though that these phenomena constitute the basis of all evolutionary science and the very central problem of natural history is admitted by all. Nor is this due to the special difficulty of such inquiries so much as to general neglect of the subject.

* The first half of this paper is reprinted with additions and modifications from the *Journal of the Royal Horticultural Society*, 1900, vol. xxv., parts 1 and 2. Written almost immediately after the rediscovery of Mendel, it will be seen to be already in some measure out of date, but it may thus serve to show the relation of the new conceptions to the old.

It is in the hope of inducing others to follow these lines of investigation that I take the problems of heredity as the subject of this lecture to the Royal Horticultural Society.

No one has better opportunities of pursuing such work than horticulturists and stock breeders. They are daily witnesses of the phenomena of heredity. Their success also depends largely on a knowledge of its laws, and obviously every increase in that knowledge is of direct and special importance to them.

The want of systematic study of heredity is due chiefly to misapprehension. It is supposed that such work requires a lifetime. But though for adequate study of the complex phenomena of inheritance long periods of time must be necessary, yet in our present state of deep ignorance almost of the outline of the facts, observations carefully planned and faithfully carried out for even a few years may produce results of great value. In fact, by far the most appreciable and definite additions to our knowledge of these matters have been thus obtained.

There is besides some misapprehension as to the kind of knowledge which is especially wanted at this time, and as to the modes by which we may expect to obtain it. The present paper is written in the hope that it may in some degree help to clear the ground of these difficulties by a preliminary consideration of the question, How far have we got towards an exact knowledge of heredity, and how can we get further?

Now this is pre-eminently a subject in which we must distinguish what we *can* do from what we want to do. We *want* to know the whole truth of the matter; we want to know the physical basis, the inward and essential nature, "the causes," as they are sometimes called, of heredity: but we want also to know the laws which the outward and visible phenomena obey.

Let us recognise from the outset that as to the essential nature of these phenomena we still know absolutely nothing. We have no glimmering of an idea as to what constitutes the essential process by which the likeness of the parent is transmitted to the offspring. We can study the processes of fertilisation and development in the finest detail which the microscope manifests to us, and we may fairly say that we have now a considerable grasp of the visible phenomena; but of the nature of the physical basis of heredity we have no conception at all. No one has yet any suggestion, working hypothesis, or mental picture that has thus far helped in the slightest degree to penetrate beyond what we see. The process is as utterly mysterious to us as a flash of lightning is to a savage. We do not know what is the essential agent in the transmission of parental characters, not even whether it is a material agent or not. Not only is our ignorance complete, but no one has the remotest idea how to set to work on that part of the problem. We are in the state in which the students of physical science were, in the period when it was open to anyone to believe that heat was a material substance or not, as he chose.

But apart from any conception of the essential modes of transmission of characters, we *can* study the outward facts of the transmission. Here, if our knowledge is still very vague, we are at least beginning to see how we ought to go to work. Formerly naturalists were content with the collection of numbers of isolated instances of transmission—more especially, striking and peculiar cases—the sudden appearance of highly prepotent forms, and the like. We are now passing out of that stage. It is not that the interest of particular cases has in any way diminished—for such records will always have their value—but it has become likely that general expressions will be found capable of sufficiently wide application to be justly called "laws" of heredity. That this is so was till recently due almost entirely to the work of Mr F. Galton, to whom we are indebted for the first systematic attempt to enuntiate such a law.

All laws of heredity so far propounded are of a statistical character and have been obtained by statistical methods. If we consider for a moment what is actually meant by a "law of heredity" we shall see at once why these investigations must follow statistical methods. For a "law" of heredity is simply an attempt to declare the course of heredity under given conditions. But if we attempt to predicate the course of heredity we have to deal with conditions and groups of causes wholly unknown to us, whose presence we cannot recognize, and whose magnitude we cannot estimate in any particular case. The course of heredity in particular cases therefore cannot be foreseen.

Of the many factors which determine the degree to which a given character shall be present in a given individual only one is usually known to us, namely, the degree to which that character is present in the parents. It is common knowledge that there is not that close correspondence between parent and offspring which would result were this factor the only one operating; but that, on the contrary, the resemblance between the two is only an uncertain one.

In dealing with phenomena of this class the study

of single instances reveals no regularity. It is only by collection of facts in great numbers, and by statistical treatment of the mass, that any order or law can be perceived. In the case of a chemical reaction, for instance, by suitable means the conditions can be accurately reproduced, so that in every individual case we can predict with certainty that the same result will occur. But with heredity it is somewhat as it is in the case of the rainfall. No one can say how much rain will fall to-morrow in a given place, but we can predict with moderate accuracy how much will fall next year, and for a period of years a prediction can be made which accords very closely with the truth.

Similar predictions can from statistical data be made as to the duration of life and a great variety of events, the conditioning causes of which are very imperfectly understood. It is predictions of this kind that the study of heredity is beginning to make possible, and in that sense laws of heredity can be perceived.

We are as far as ever from knowing *why* some characters are transmitted, while others are not; nor can anyone yet foretell which individual parent will transmit characters to the offspring, and which will not; nevertheless the progress made is distinct.

As yet investigations of this kind have been made in only a few instances, the most notable being those of Galton on human stature, and on the transmission of colours in Basset hounds. In each of these cases he has shown that the expectation of inheritance is such that a simple arithmetical rule is approximately followed. The rule thus arrived at is that of the whole heritage of the offspring the two parents together on an average contribute one half, the four grandparents one-quarter, the eight great-grandparents one-eighth, and so on, the remainder being contributed by the remoter ancestors.

Such a law is obviously of practical importance. In any case to which it applies we ought thus to be able to predict the degree with which the purity of a strain may be increased by selection in each successive generation.

To take a perhaps impossibly crude example, if a seedling show any particular character which it is desired to fix, on the assumption that successive self-fertilisations are possible, according to Galton's law the expectation of purity should be in the first generation of self-fertilisation 1 in 2, in the second generation 3 in 4, in the third 7 in 8, and so on*.

But already many cases are known to which the rule in any simple form will not apply. Galton points out that it takes no account of individual prepotencies. There are, besides, numerous cases in which on crossing two varieties the character of one variety almost always appears in each member of the first cross-bred generation. Examples of these will be familiar to those who have experience in such matters. The offspring of the Polled Angus cow and the Shorthorn bull is almost invariably polled or with very small loose "scurs." Seedlings raised by crossing *Atropa belladonna* with the yellow-fruited variety have without exception the blackish-purple fruits of the type. In several hairy species when a cross with a glabrous variety is made, the first cross-bred generation is altogether hairy†.

Still more numerous are examples in which the characters of one variety very largely, though not exclusively, predominate in the offspring.

* See later. Galton gave a simple diagrammatic representation of his law in *Nature*, 1898, vol. LVII. p. 293.

† These we now recognize as examples of Mendelian 'dominance.'

These large classes of exceptions—to go no further—indicate that, as we might in any case expect, the principle is not of universal application, and will need various modifications if it is to be extended to more complex cases of inheritance of varietal characters. No more useful work can be imagined than a systematic determination of the precise "law of heredity" in numbers of particular cases.

Until lately the work which Galton accomplished stood almost alone in this field, but quite recently remarkable additions to our knowledge of these questions have been made. In the year 1900 Professor de Vries published a brief account* of experiments which he has for several years been carrying on, giving results of the highest value.

The description is very short, and there are several points as to which more precise information is necessary both as to details of procedure and as to statement of results. Nevertheless it is impossible to doubt that the work as a whole constitutes a marked step forward, and the full publication which is promised will be awaited with great interest.

The work relates to the course of heredity in cases where definite varieties differing from each other in some *one* definite character are crossed together. The cases are all examples of discontinuous variation: that is to say, cases in which actual intermediates between the parent forms are not usually produced on crossing†. It is shown that the subsequent posterity obtained by self-fertilising these cross-breds or hybrids, or by breeding them with each other, break up into the original parent forms according to fixed numerical rule.

* *Comptes Rendus*, March 26, 1900, and *Ber. d. Deutsch. Bot. Ges.* xviii. 1900, p. 83.

† This conception of discontinuity is of course pre-Mendelian.

Professor de Vries begins by reference to a remarkable memoir by Gregor Mendel*, giving the results of his experiments in crossing varieties of *Pisum sativum.* These experiments of Mendel's were carried out on a large scale, his account of them is excellent and complete, and the principles which he was able to deduce from them will certainly play a conspicuous part in all future discussions of evolutionary problems. It is not a little remarkable that Mendel's work should have escaped notice, and been so long forgotten.

For the purposes of his experiments Mendel selected seven pairs of characters as follows:—

1. Shape of ripe seed, whether round; or angular and wrinkled.

2. Colour of "endosperm" (cotyledons), whether some shade of yellow; or a more or less intense green.

3. Colour of the seed-skin, whether various shades of grey and grey-brown; or white.

4. Shape of seed-pod, whether simply inflated; or deeply constricted between the seeds.

5. Colour of unripe pod, whether a shade of green; or bright yellow.

6. Nature of inflorescence, whether the flowers are arranged along the axis of the plant; or are terminal and form a kind of umbel.

7. Length of stem, whether about 6 or 7 ft. long, or about $\frac{3}{4}$ to $1\frac{1}{2}$ ft.

Large numbers of crosses were made between Peas differing in respect of *one* of each of these pairs of characters.

* 'Versuche üb. Pflanzenhybriden' in the *Verh. d. Naturf. Ver. Brünn*, iv. 1865.

It was found that in each case the offspring of the cross exhibited the character of one of the parents in almost undiminished intensity, and intermediates which could not be at once referred to one or other of the parental forms were not found.

In the case of each pair of characters there is thus one which in the first cross prevails to the exclusion of the other. This prevailing character Mendel calls the *dominant* character, the other being the *recessive* character*.

That the existence of such "dominant" and "recessive" characters is a frequent phenomenon in cross-breeding, is well known to all who have attended to these subjects.

By letting the cross-breds fertilise themselves Mendel next raised another generation. In this generation were individuals which showed the dominant character, but also individuals which presented the recessive character. Such a fact also was known in a good many instances. But Mendel discovered that in this generation the numerical proportion of dominants to recessives is on an average of cases approximately constant, being in fact *as three to one.* With very considerable regularity these numbers were approached in the case of each of his pairs of characters.

There are thus in the first generation raised from the cross-breds 75 per cent. dominants and 25 per cent. recessives.

These plants were again self-fertilised, and the offspring of each plant separately sown. It next appeared that the offspring of the recessives *remained pure recessive,* and in subsequent generations never produced the dominant again.

But when the seeds obtained by self-fertilising the

* Note that by these novel terms the complications involved by use of the expression "prepotent" are avoided.

dominants were examined and sown it was found that the dominants were not all alike, but consisted of two classes, (1) those which gave rise to pure dominants, and (2) others which gave a mixed offspring, composed partly of recessives, partly of dominants. Here also it was found that the average numerical proportions were constant, those with pure dominant offspring being to those with mixed offspring as one to two. Hence it is seen that the 75 per cent. dominants are not really of similar constitution, but consist of twenty-five which are pure dominants and fifty which are really cross-breds, though, like the cross-breds raised by crossing the two original varieties, they only exhibit the dominant character.

To resume, then, it was found that by self-fertilising the original cross-breds the same proportion was always approached, namely—

25 dominants, 50 cross-breds, 25 recessives,

or $1D : 2DR : 1R$.

Like the pure recessives, the pure dominants are thenceforth pure, and only give rise to dominants in all succeeding generations studied.

On the contrary the fifty cross-breds, as stated above, have mixed offspring. But these offspring, again, in their numerical proportions, follow the same law, namely, that there are three dominants to one recessive. The recessives are pure like those of the last generation, but the dominants can, by further self-fertilisation, and examination or cultivation of the seeds produced, be again shown to be made up of pure dominants and cross-breds in the same proportion of one dominant to two cross-breds.

The process of breaking up into the parent forms is thus continued in each successive generation, the same numerical law being followed so far as has yet been observed.

Mendel made further experiments with *Pisum sativum*, crossing pairs of varieties which differed from each other in *two* characters, and the results, though necessarily much more complex, showed that the law exhibited in the simpler case of pairs differing in respect of one character operated here also.

In the case of the union of varieties AB and ab differing in two distinct pairs of characters, A and a, B and b, of which A and B are dominant, a and b recessive, Mendel found that in the first cross-bred generation there was only *one* class of offspring, really $AaBb$.

But by reason of the dominance of one character of each pair these first crosses were hardly if at all distinguishable from AB.

By letting these $AaBb$'s fertilise themselves, only *four* classes of offspring seemed to be produced, namely,

AB	showing	both dominant characters.
Ab	,,	dominant A and recessive b.
aB	,,	recessive a and dominant B.
ab	,,	both recessive characters a and b.

The numerical ratio in which these classes appeared were also regular and approached the ratio

$$9AB : 3Ab : 3aB : 1ab.$$

But on cultivating these plants and allowing them to fertilise themselves it was found that the members of the

Ratios		
1		ab class produce only ab's.
3	1	aB class may produce either all aB's,
	2	*or* both aB's and ab's.

RATIOS

3	1	*Ab* class may produce either all *Ab*'s,
	2	*or* both *Ab*'s and *ab*'s.
9	1	*AB* class may produce either all *AB*'s,
	2	*or* both *AB*'s and *Ab*'s,
	2	*or* both *AB*'s and *aB*'s,
	4	*or* all four possible classes again, namely, *AB*'s, *Ab*'s, *aB*'s, and *ab*'s,

and the average number of members of each class will approach the ratio 1 : 3 : 3 : 9 as indicated above.

The details of these experiments and of others like them made with *three* pairs of differentiating characters are all set out in Mendel's memoir.

Professor de Vries has worked at the same problem in some dozen species belonging to several genera, using pairs of varieties characterised by a great number of characters: for instance, colour of flowers, stems, or fruits, hairiness, length of style, and so forth. He states that in all these cases Mendel's principles are followed.

The numbers with which Mendel worked, though large, were not large enough to give really smooth results*; but with a few rather marked exceptions the observations are remarkably consistent, and the approximation to the numbers demanded by the law is greatest in those cases where the largest numbers were used. When we consider, besides, that Tschermak and Correns announce definite confirmation in the case of *Pisum*, and de Vries adds the evidence of his long series of observations on other species and orders, there can be no doubt that Mendel's law is a substantial

* Professor Weldon (p. 232) takes great exception to this statement, which he considerately attributes to "some writers." After examining the conclusions he obtained by algebraical study of Mendel's figures I am disposed to think my statement not very far out.

reality; though whether some of the cases that depart most widely from it can be brought within the terms of the same principle or not, can only be decided by further experiments.

One may naturally ask, How can these results be brought into harmony with the facts of hybridisation hitherto known; and, if all this is true, how is it that others who have carefully studied the phenomena of hybridisation have not long ago perceived this law? The answer to this question is given by Mendel at some length, and it is, I think, satisfactory. He admits from the first that there are undoubtedly cases of hybrids and cross-breds which maintain themselves pure and do not break up. Such examples are plainly outside the scope of his law. Next he points out, what to anyone who has rightly comprehended the nature of discontinuity in variation is well known, that the variations in *each* character must be *separately* regarded. In most experiments in crossing, forms are taken which differ from each other in a multitude of characters—some continuous, others discontinuous, some capable of blending with their contraries, while others are not. The observer on attempting to perceive any regularity is confused by the complications thus introduced. Mendel's law, as he fairly says, could only appear in such cases by the use of overwhelming numbers, which are beyond the possibilities of practical experiment. Lastly, no previous observer had applied a strict statistical method.

Both these answers should be acceptable to those who have studied the facts of variation and have appreciated the nature of Species in the light of those facts. That different species should follow different laws, and that the same law should not apply to all characters alike, is exactly what we have every right to expect. It will also be remembered that the principle is only explicitly declared to apply to discontinuous characters*. As stated also it can only be true where reciprocal crossings lead to the same result. Moreover, it can only be tested when there is no sensible diminution in fertility on crossing.

Upon the appearance of de Vries' paper announcing the "rediscovery" and confirmation of Mendel's law and its extension to a great number of cases two other observers came forward almost simultaneously and independently described series of experiments fully confirming Mendel's work. Of these papers the first is that of Correns, who repeated Mendel's original experiment with Peas having seeds of different colours. The second is a long and very valuable memoir of Tschermak, which gives an account of elaborate researches into the results of crossing a number of varieties of *Pisum sativum*. These experiments were in many cases carried out on a large scale, and prove the main fact enuntiated by Mendel beyond any possibility of contradiction. The more exhaustive of these researches are those of Tschermak on Peas and Correns on several varieties of Maize. Both these elaborate investigations have abundantly proved the general applicability of Mendel's law to the character of the plants studied, though both indicate some few exceptions. The details of de Vries' experiments are promised in the second volume of his most valuable *Mutationstheorie*. Correns in regard to Maize and Tschermak in the case of *P. sativum* have obtained further proof that Mendel's law holds as well in the case of varieties differing from each other in *two* pairs of characters, one of each pair being dominant, though of course a more complicated expression is needed in such cases†.

* See later.

† Tschermak's investigations were besides directed to a re-exami-

That we are in the presence of a new principle of the highest importance is manifest. To what further conclusions it may lead us cannot yet be foretold. But both Mendel and the authors who have followed him lay stress on one conclusion, which will at once suggest itself to anyone who reflects on the facts. For it will be seen that the results are such as we might expect if it be imagined that the cross-bred plant produced pollen grains and egg-cells, each of which bears only *one* of the alternative varietal characters and not both. If this were so, and if on an average the same number of pollen grains and egg-cells transmit each of the two characters, it is clear that on a random assortment of pollen grains and egg-cells Mendel's law would be obeyed. For 25 per cent. of "dominant" pollen grains would unite with 25 per cent. "dominant" egg-cells; 25 per cent. "recessive" pollen grains would similarly unite with 25 per cent. "recessive" egg-cells; while the remaining 50 per cent. of each kind would unite together. It is this consideration which leads both Mendel and those who have followed him to assert that these facts of crossing prove that each egg-cell and each pollen grain is pure in respect of each character to which the law applies. It is highly desirable that varieties differing in the form of their pollen should be made the subject of these experiments, for it is quite possible that in such a case strong confirmation of this deduction might be obtained. [Preliminary trials made with reference to this point have so far given negative results. Remembering that a pollen grain is not a germ-cell, but only a bearer of

nation of the question of the absence of beneficial results on cross-fertilising *P. sativum*, a subject already much investigated by Darwin, and upon this matter also important further evidence is given in great detail.

a germ-cell, the hope of seeing pollen grains differentiated according to the characters they bear is probably remote. Better hopes may perhaps be entertained in regard to spermatozoa, or possibly female cells.]

As an objection to the deduction of purity of germ-cells, however, it is to be noted that though true intermediates did not generally occur, yet the intensity in which the characters appeared did vary in degree, and it is not easy to see how the hypothesis of *perfect* purity in the reproductive cells can be supported in such cases. Be this, however, as it may, there is no doubt we are beginning to get new lights of a most valuable kind on the nature of heredity and the laws which it obeys. It is to be hoped that these indications will be at once followed up by independent workers. Enough has been said to show how necessary it is that the subjects of experiment should be chosen in such a way as to bring the laws of heredity to a real test. For this purpose the first essential is that the differentiating characters should be few, and that all avoidable complications should be got rid of. Each experiment should be reduced to its simplest possible limits. The results obtained by Galton, and also the new ones especially described in this paper, have each been reached by restricting the range of observation to one character or group of characters, and it is certain that by similar treatment our knowledge of heredity may be rapidly extended.

To the above popular presentation of the essential facts, made for an audience not strictly scientific, some addition, however brief, is called for. First, in regard to the law of Ancestry, spoken of on p. 5. Those who are acquainted with Pearson's *Grammar of Science*, 2nd ed. published early in

1900, the same author's paper in *Proc. R. S.* vol. 66, 1900, p. 140, or the extensive memoir (pubd. Oct. 1900), on the inheritance of coat-colour in horses and eye-colour in man (*Phil. Trans.* 195, A, 1900, p. 79), will not need to be told that the few words I have given above constitute a most imperfect diagram of the operations of that law as now developed. Until the appearance of these treatises it was, I believe, generally considered that the law of Ancestral Heredity was to be taken as applying to phenomena like these (coat-colour, eye-colour, &c.) where the inheritance is generally *alternative*, as well as to the phenomena of *blended* inheritance.

Pearson, in the writings referred to, besides withdrawing other large categories of phenomena from the scope of its operations, points out that the law of Ancestral Heredity does not satisfactorily express the cases of alternative inheritance. He urges, and with reason, that these classes of phenomena should be separately dealt with.

The whole issue as regards the various possibilities of heredity now recognized will be made clearer by a very brief exposition of the several conceptions involved.

If an organism producing germ-cells of a given constitution, uniform in respect of the characters they bear, breeds with another organism* bearing *precisely similar* germ-cells, the offspring resulting will, if the conditions are identical, be uniform.

In practice such a phenomenon is seen in *pure*-breeding. It is true that we know no case in nature where all the germ-cells are thus identical, and where no variation takes place beyond what we can attribute to conditions, but we

* For simplicity the case of self-fertilisation is omitted from this consideration.

know many cases where such a result is approached, and very many where all the essential features which we regard as constituting the characters of the breed are reproduced with approximate certainty in every member of the pure-bred race, which thus closely approach to uniformity.

But if two germ-cells of dissimilar constitution unite in fertilisation, what offspring are we to expect* ? First let us premise that the answer to this question is known experimentally to differ for many organisms and for many classes of characters, and may almost certainly be in part determined by external circumstances. But omitting the last qualification, certain principles are now clearly detected, though what principle will apply in any given case can only be determined by direct experiment made with that case.

This is the phenomenon of *cross*-breeding. As generally used, this term means the union of members of dissimilar varieties, or species: though when dissimilar gametes† produced by two individuals of the same variety unite in fertilisation, we have essentially *cross*-breeding in respect of the character or characters in which those gametes differ. We will suppose, as before, that these two gametes bearing properties unlike in respect of a given character, are borne by different individuals.

In the simplest case, suppose a gamete from an individual presenting any character in intensity A unite in fertilisation with another from an individual presenting the same character in intensity a. For brevity's sake we

* In all the cases discussed it is assumed that the gametes are similar except in regard to the "heritage" they bear, and that no *original* variation is taking place. The case of mosaics is also left wholly out of account (see later).

† The term "gamete" is now generally used as the equivalent of "germ-cell," whether male or female, and the term "zygote" is here used for brevity to denote the organism resulting from fertilisation.

may call the parent individuals *A* and *a*, and the resulting zygote *Aa*. What will the structure of *Aa* be in regard to the character we are considering?

Up to Mendel no one proposed to answer this question in any other way than by reference to the intensity of the character in the progenitors, and *primarily* in the parents, *A* and *a*, in whose bodies the gametes had been developed. It was well known that such a reference gave a very poor indication of what *Aa* would be. Both *A* and *a* may come from a population consisting of individuals manifesting the same character in various intensities. In the pedigree of either *A* or *a* these various intensities may have occurred few or many times. Common experience leads us to expect the probability in regard to *Aa* to be influenced by this history. The next step is that which Galton took. He extended the reference beyond the immediate parents of *Aa*, to its grandparents, great-grandparents, and so on, and in the cases he studied he found that from a knowledge of the intensity in which the given character was manifested in each progenitor, even for some few generations back, a fairly accurate prediction could be made, not as to the character of any individual *Aa*, but as to the average character of *Aa*'s of similar parentage, in general.

But suppose that instead of individuals presenting one character in differing intensities, two individuals breed together distinguished by characters which we know to be mutually exclusive, such as *A* and *B*. Here again we may speak of the individuals producing the gametes as *A* and *B*, and the resulting zygote as *AB*. What will *AB* be like? The population here again may consist of many like *A* and like *B*. These two forms may have been breeding together indiscriminately, and there may have been many or few of either type in the pedigree of either *A* or *B*.

Here again Galton applied his method with remarkable success. Referring to the progenitors of A and B, determining how many of each type there were in the direct pedigree of A and of B, he arrived at the same formula as before, with the simple difference that instead of expressing the probable average intensity of one character in several individuals, the prediction is given in terms of the probable number of A's and B's that would result on an average when particular A's and B's of known pedigree breed together.

The law as Galton gives it is as follows:—

"It is that the two parents contribute between them on the average one-half, or (0·5) of the total heritage of the offspring; the four grandparents, one-quarter, or $(0{\cdot}5)^2$; the eight great-grandparents, one-eighth, or $(0{\cdot}5)^3$, and so on. Then the sum of the ancestral contributions is expressed by the series

$$\{(0{\cdot}5)+(0{\cdot}5)^2+(0{\cdot}5)^3, \text{\&c.}\},$$

which, being equal to 1, accounts for the whole heritage."

In the former case where A and a are characters which can be denoted by reference to a common scale, the law assumes of course that the inheritance will be, to use Galton's term, *blended*, namely that the zygote resulting from the union of A with a will on the average be more like a than if A had been united with A; and conversely that an Aa zygote will on the average *be more like A than an aa zygote would be.*

But in the case of A's and B's, which are assumed to be mutually exclusive characters, we cannot speak of blending, but rather, to use Galton's term, of *alternative* inheritance.

Pearson, finding that the law whether formulated thus,

or in the modified form in which he restated it*, did not express the phenomena of alternative inheritance known to him with sufficient accuracy to justify its strict application to them, and also on general grounds, proposed that the phenomena of blended and alternative inheritance should be treated apart—a suggestion† the wisdom of which can scarcely be questioned.

Now the law thus imperfectly set forth and every modification of it is incomplete in one respect. It deals only with the characters of the resulting zygotes and predicates nothing in regard to the gametes which go to form them. A good prediction may be made as to any given group of zygotes, but the various possible constitutions of the gametes are not explicitly treated.

Nevertheless a definite assumption is implicitly made regarding the gametes. It is not in question that differences between these gametes may occur in respect of the heritage they bear; yet it is assumed that these differences will be distributed among the gametes of any individual zygote in such a way that each gamete remains capable, on fertilisation, of transmitting *all* the characters (both of the parent-zygote and of its progenitors) to the zygote which it then contributes to form (and to the posterity of that zygote) in the intensity indicated by the law. Hence the gametes of any individual are taken as collectively a fair sample of all the racial characters in their appropriate intensities, and this theory demands that there shall have been no qualitative redistribution of characters among the gametes of any zygote in such a way that some gametes shall be finally excluded from partaking of and transmitting any specific

* In Pearson's modification the parents contribute 0·3, the grandparents 0·15, the great-grandparents ·075.

† See the works referred to above.

part of the heritage. The theory further demands—and by the analogy of what we know otherwise not only of animals and plants, but of physical or chemical laws, perhaps this is the most serious assumption of all—that the structure of the gametes shall admit of their being capable of transmitting any character in any intensity varying from zero to totality with equal ease; and that gametes of each intensity are all equally likely to occur, given a pedigree of appropriate arithmetical composition.

Such an assumption appears so improbable that even in cases where the facts seem as yet to point to this conclusion with exceptional clearness, as in the case of human stature, I cannot but feel there is still room for reserve of judgment.

However this may be, the Law of Ancestral Heredity, and all modifications of it yet proposed, falls short in the respect specified above, that *it does not directly attempt to give any account of the distribution of the heritage among the gametes* of any one individual.

Mendel's conception differs fundamentally from that involved in the Law of Ancestral Heredity. The relation of his hypothesis to the foregoing may be most easily shown if we consider it first in application to the phenomena resulting from the cross-breeding of two pure varieties.

Let us again consider the case of two varieties each displaying the same character, but in the respective intensities A and a. Each gamete of the A variety bears A, and each gamete of the a variety bears a. When they unite in fertilisation they form the zygote Aa. What will be its characters? The Mendelian teaching would reply that this can only be known by direct experiment with the two forms A and a, and that the characters A and a perceived in those two forms or varieties need not give any indication as to the character of the zygote *Aa*. It may display the character *A*, or *a*, or a character half way between the two, or a character beyond *A* or below *a*. The character of *Aa* is not regarded as a *heritage* transmitted to it by *A* and by *a*, but as a character special and peculiar to *Aa*, just as NaCl is not a body half way between sodium and chlorine, or such that its properties can be predicted from or easily stated in terms of theirs.

If a concrete case may help, a tall pea *A* crossed with a dwarf *a* often produces, not a plant having the height of either *A* or *a*, but something *taller* than the pure tall variety *A*.

But if the case obeys the Mendelian principles—as does that here quoted—then it can be declared *first* that the gametes of *Aa* will not be bearers of the character proper to *Aa*; but, generally speaking, each gamete will either bear the pure *A* character or the pure *a* character. There will in fact be a redistribution of the characters brought in by the gametes which united to form the zygote *Aa*, such that each gamete of *Aa* is pure, as the parental gametes were. *Secondly* this redistribution will occur in such a way that, of the gametes produced by such *Aa*'s, on an average there will be equal numbers of *A* gametes and of *a* gametes.

Consequently if *Aa*'s breed together, the new *A* gametes may meet each other in fertilisation, forming a zygote *AA*, namely, the pure *A* variety again; similarly two *a* gametes may meet and form *aa*, or the pure *a* variety again. But if an *A* gamete meets an *a* it will once more form *Aa*, with its special character. This *Aa* is the hybrid, or "mule" form, or as I have elsewhere called it, the *heterozygote*, as distinguished from *AA* or *aa* the *homozygotes*.

Similarly if the two gametes of two varieties distinguished by characters, A and B, which cannot be described in terms of any common scale (such as for example the "rose" and "single" combs of fowls) unite in fertilisation, again the character of the mule form cannot be predicted. Before the experiment is made the "mule" may present *any* form. Its character or properties can as yet be no more predicted than could those of the compounds of unknown elements before the discovery of the periodic law.

But again—if the case be Mendelian—the gametes borne by AB will be either A's or B's*, and the cross-bred AB's breeding together will form AA's, AB's and BB's. Moreover, if as in the normal Mendelian case, AB's bear on an average equal numbers of A gametes and B gametes, the numerical ratio of these resulting zygotes to each other will be

$$1\,AA : 2\,AB : 1\,BB.$$

We have seen that Mendel makes no prediction as to the outward and visible characters of AB, but only as to the essential constitution and statistical condition of its gametes in regard to the characters A and B. Nevertheless in a large number of cases the character of AB is known to fall into one of three categories (omitting mosaics).

(1) The cross-bred may almost always resemble one of its pure parents so closely as to be practically indistinguishable from that pure form, as in the case of the yellow cotyledon-colour of certain varieties of peas when crossed with green-cotyledoned varieties; in which case the parental character, yellow, thus

* This conception was clearly formed by Naudin simultaneously with Mendel, but it was not worked out by him and remained a mere suggestion. In one place also Focke came very near to the same idea (see Bibliography).

manifested by the cross-bred is called "dominant" and the parental character, green, not manifested, is called recessive.

(2) The cross-bred may present some condition intermediate between the two parental forms, in which case we may still retain the term "blend" as applied to the zygote.

Such an "intermediate" may be the apparent mean between the two parental forms or be nearer to one or other in any degree. Such a case is that of a cross between a rich crimson Magenta Chinese Primrose and a clear White, giving a flower of a colour appropriately described as a "washy" magenta.

(3) The cross-bred may present some form quite different from that of either pure parent. Though, as has been stated, nothing can be predicted of an unknown case, we already know a considerable number of examples of this nature in which the mule-form *approaches sometimes with great accuracy to that of a putative ancestor, near or remote.* It is scarcely possible to doubt that several—though perhaps not all—of Darwin's "reversions on crossing" were of this nature.

Such a case is that of the "wild grey mouse" produced by the union of an albino tame mouse and a piebald Japanese mouse*. These "reversionary" mice bred together produce the parental tame types, some other types, and "reversionary" mice again.

From what has been said it will now be clear that the applicability of the Mendelian hypothesis has, intrinsically,

* See von Guaita, *Ber. naturf. Ges. Freiburg* x. 1898 and xi. 1899, quoted by Professor Weldon (see later).

nothing whatever to do with the question of the inheritance being *blended* or *alternative*. In fact, as soon as the relation of zygote characters to gamete characters is appreciated, it is difficult to see any reason for supposing that the manifestation of characters seen in the zygotes should give any indication as to their mode of allotment among the gametes.

On a previous occasion I pointed out that the terms "Heredity" and "Inheritance" are founded on a misapplication of metaphor, and in the light of our present knowledge it is becoming clearer that the ideas of "transmission" of a character by parent to offspring, or of there being any "contribution" made by an ancestor to its posterity, must only be admitted under the strictest reserve, and merely as descriptive terms.

We are now presented with some entirely new conceptions :—

(1) The purity of the gametes in regard to certain characters.

(2) The distinction of all zygotes according as they are or are not formed by the union of like or unlike gametes. In the former case, apart from Variation, they breed true when mated with their like; in the latter case their offspring, collectively, will be heterogeneous.

(3) If the zygote be formed by the union of dissimilar gametes, we may meet the phenomenon of (*a*) dominant and recessive characters; (*b*) a blend form; (*c*) a form distinct from either parent, often reversionary*.

* This fact sufficiently indicates the difficulties involved in a superficial treatment of the phenomenon of reversion. To call such reversions as those named above "returns to ancestral type" would be, if more than a descriptive phrase were intended, quite misleading.

But there are additional and even more significant deductions from the facts. We have seen that the gametes are differentiated in respect of pure characters. Of these pure characters there may *conceivably* be any number associated together in one organism. In the pea Mendel detected at least seven—not all seen by him combined in the same plant, but there is every likelihood that they are all capable of being thus combined.

Each such character, which is capable of being dissociated or replaced by its contrary, must henceforth be conceived of as a distinct *unit-character*; and as we know that the several unit-characters are of such a nature that any one of them is capable of independently displacing or being displaced by one or more alternative characters taken singly, we may recognize this fact by naming such unit-characters *allelomorphs*. So far, we know very little of any allelomorphs existing otherwise than as *pairs* of contraries, but this is probably merely due to experimental limitations and the rudimentary state of our knowledge.

In one case (combs of fowls) we know three characters, *pea* comb, *rose* comb and *single* comb; of which *pea* and *single*, or *rose* and *single*, behave towards each other as a pair of allelomorphs, but of the behaviour of *pea* and *rose* towards each other we know as yet nothing.

We have no reason as yet for affirming that any phenomenon properly described as *displacement* of one allelomorph by another occurs, though the metaphor may be a useful one. In all cases where *dominance* has been perceived, we can affirm that the members of the allelomorphic pair stand to each other in a relation the nature

It is not the ancestral *type* that has come back, but something else has come in its guise, as the offspring presently prove. For the first time we thus begin to get a rationale of "reversion."

of which we are as yet wholly unable to apprehend or illustrate.

To the new conceptions already enumerated we may therefore add

(4) *Unit-characters* of which some, *when once arisen by Variation*, are alternative to each other in the constitution of the gametes, according to a definite system.

From the relations subsisting between these characters, it follows that as each zygotic union of allelomorphs is *resolved* on the formation of the gametes, no zygote can give rise to gametes collectively representing more than *two* characters allelomorphic to each other, apart from new variation.

From the fact of the existence of the interchangeable characters we must, for purposes of treatment, and to complete the possibilities, necessarily form the conception of an *irresoluble base*, though whether such a conception has any objective reality we have no means as yet of determining.

We have now seen that when the varieties A and B are crossed together, the heterozygote, AB, produces gametes bearing the pure A character and the pure B character. In such a case we speak of such characters as *simple* allelomorphs. In many cases however a more complex phenomenon happens. The character brought in on fertilisation by one or other parent may be of such a nature that when the zygote, AB, forms its gametes, these are not individually bearers merely of A and B, *but of a number of characters themselves again integral*, which in, say A, behaved as one character so long as its gametes united in fertilisation with others like themselves, but on cross-fertilisation are resolved and redistributed among the gametes produced by the cross-bred zygote.

In such a case we call the character A a *compound*

allelomorph, and we can speak of the integral characters which constitute it as *hypallelomorphs.* We ought to write the heterozygote $(AA'A''...)\ B$ and the gametes produced by it may be of the form A, A', A'', A''',...B. Or the resolution may be incomplete in various degrees, as we already suspect from certain instances; in which case we may have gametes A, $A'A''$, $A'''A''''$, $A'A''A^{v}$,...B, and so on. Each of these may meet a similar or a dissimilar gamete in fertilisation, forming either a homozygote, or a heterozygote with its distinct properties.

In the case of compound allelomorphs we know as yet nothing of the statistical relations of the several gametes.

Thus we have the conception

(5) *of a Compound character*, borne by one gamete, transmitted entire as a single character so long as fertilisation only occurs between like gametes, or is, in other words, "symmetrical," but if fertilisation take place with a dissimilar gamete (or possibly by other causes), resolved into integral constituent-characters, each separately transmissible.

Next, as, by the union of the gametes bearing the various hypallelomorphs with other such gametes, or with gametes bearing simple allelomorphs, in fertilisation, a number of new zygotes will be formed, such as may not have been seen before in the breed: these will inevitably be spoken of as *varieties*; and it is difficult not to extend the idea of variation to them. To distinguish these from other variations—which there must surely be—we may call them

(6) *Analytical* variations in contradistinction to

(7) *Synthetical* variations, occurring not by the separation of pre-existing constituent-characters but by the addition of new characters.

Lastly, it is impossible to be presented with the fact that in Mendelian cases the cross-bred produces on an average *equal* numbers of gametes of each kind, that is to say, a symmetrical result, without suspecting that this fact must correspond with some symmetrical figure of distribution of those gametes in the cell-divisions by which they are produced.

At the present time these are the main conceptions—though by no means all—arising directly from Mendel's work. The first six are all more or less clearly embodied by him, though not in every case developed in accordance with modern knowledge. The seventh is not a Mendelian conception, but the facts before us justify its inclusion in the above list though for the present it is little more than a mere surmise.

In Mendelian cases it will now be perceived that all the zygotes composing the population consist of a limited number of possible types, each of definite constitution, bearing gametes also of a limited and definite number of types, and definite constitution in respect of pre-existing characters. It is now evident that in such cases each several progenitor need not be brought to account in reckoning the probable characters of each descendant; for the gametes of cross-breds are differentiated at each successive generation, some parental (Mendelian) characters being left out in the composition of each gamete produced by a zygote arising by the union of bearers of opposite allelomorphs.

When from these considerations we return to the phenomena comprised in the Law of Ancestral Heredity, what certainty have we that the same conceptions are not applicable there also?

It has now been shown that the question whether in the cross-bred zygotes in general the characters blend or are mutually exclusive is an entirely subordinate one, and distinctions with regard to the essential nature of heredity based on these circumstances become irrelevant.

In the case of a population presenting continuous variation in regard to say, stature, it is easy to see how purity of the gametes in respect of any intensities of that character might not in ordinary circumstances be capable of detection. There are doubtless more than two pure gametic forms of this character, but there may quite conceivably be six or eight. When it is remembered that each heterozygous combination of any two may have its own appropriate stature, and that such a character is distinctly dependent on external conditions, the mere fact that the observed curves of stature give "chance distributions" is not surprising and may still be compatible with purity of gametes in respect of certain pure types. In peas (*P. sativum*), for example, from Mendel's work we know that the tall forms and the extreme dwarf forms exhibit gametic purity. I have seen at Messrs Sutton's strong evidence of the same nature in the case of the tall Sweet Pea (*Lathyrus odoratus*) and the dwarf or procumbent "Cupid" form.

But in the case of the Sweet Pea we know at least one pure form of definitely intermediate height, and in the case of *P. sativum* there are many. When the *extreme* types breed together it will be remembered the heterozygote commonly exceeds the taller in height. In the next generation, since there is, in the case of extremes, so much margin between the types of the two pure forms, the return of the offspring to the three forms of which two are homozygous and one heterozygous is clearly perceptible.

If however instead of pure extreme varieties we were to take a pair of varieties differing normally by only a foot or two, we might, owing to the masking effects of conditions, &c., have great difficulty in distinguishing the three forms in the second generation. There would besides be twice as many heterozygous individuals as homozygous individuals of each kind, giving a symmetrical distribution of heights, and who might not—in pre-Mendelian days—have accepted such evidence—made still less clear by influence of conditions—as proof of Continuous Variation both of zygotes and gametes?

Suppose, then, that instead of two pure types, we had six or eight breeding together, each pair forming their own heterozygote, there would be a very remote chance of such purity or fixity of type whether of gamete or zygote being detected.

Dominance, as we have seen, is merely a phenomenon incidental to specific cases, between which no other common property has yet been perceived. In the phenomena of *blended* inheritance we clearly have no dominance. In the cases of *alternative* inheritance studied by Galton and Pearson there is evidently no *universal* dominance. From the tables of Basset hound pedigrees there is clearly no definite dominance of either of the coat-colours. In the case of eye-colour the published tables do not, so far as I have discovered, furnish the material for a decision, though it is scarcely possible the phenomenon, even if only occasional, could have been overlooked. We must take it, then, there is no sensible dominance in these cases; but whether there is or is not sensible gametic purity is an altogether different question, which, so far as I can judge, is as yet untouched. It may perfectly well be that we shall be compelled to recognize that in many cases there is no such purity, and that the characters may be carried by the gametes in any proportion from zero to totality, just as some substances may be carried in a solution in any proportion from zero to saturation without discontinuous change of properties. That this will be found true in *some* cases is, on any hypothesis, certain; but to prove the fact for any given case will be an exceedingly difficult operation, and I scarcely think it has been yet carried through in such a way as to leave no room for doubt.

Conversely, the *absolute* and *universal* purity of the gametes has certainly not yet been determined for any case; not even in those cases where it looks most likely that such universal purity exists. Impairment of such purity we may conceive either to occur in the form of mosaic gametes, or of gametes with blended properties. On analogy and from direct evidence we have every right to believe that gametes of both these classes may occur in rare and exceptional cases, of as yet unexplored nature*, but such a phenomenon will not diminish the significance of observed purity.

We have now seen the essential nature of the Mendelian principles and are able to appreciate the exact relation in which they stand to the group of cases included in the Law of Ancestral Heredity. In seeking any general indication as to the common properties of the phenomena which are already known to obey Mendelian principles we can as yet point to none, and whether some such common features exist or not is unknown.

There is however one group of cases, definite though as yet not numerous, where we know that the Mendelian

* It will be understood from what follows, that the existence of mosaic zygotes is no *proof* that either component gamete was mosaic.

principles do not apply. These are the phenomena upon which Mendel touches in his brief paper on *Hieracium*. As he there states, the hybrids, if they are fertile at all, produce offspring like themselves, not like their parents. In further illustration of this phenomenon he cites Wichura's *Salix* hybrids. Perhaps some dozen other such illustrations could be given which rest on good evidence. To these cases the Mendelian principle will in nowise apply, nor is it easy to conceive any modification of the law of ancestral heredity which can express them. There the matter at present rests. Among these cases, however, we perceive several more or less common features. They are often, though not always, hybrids between forms differing in many characters. The first cross frequently is not the exact intermediate between the two parental types, but may as in the few *Hieracium* cases be irregular in this respect. There is often some degree of sterility. In the absence of fuller and statistical knowledge of such cases further discussion is impossible.

Another class of cases, untouched by any hypothesis of heredity yet propounded, is that of the false hybrids of Millardet, where we have fertilisation without transmission of one or several parental characters. In these not only does the first cross show, in some respect, the character or characters of *one parent only*, but in its posterity *no re-appearance of the lost character or characters is observed.* The nature of such cases is still quite obscure, but we have to suppose that the allelomorph of one gamete only developes after fertilisation to the exclusion of the corresponding allelomorph of the other gamete, much—if the crudity of the comparison may be pardoned—as occurs on the female side in parthenogenesis without fertilisation at all.

To these as yet altogether unconformable cases we can scarcely doubt that further experiment will add many more. Indeed we already have tolerably clear evidence that many phenomena of inheritance are of a much higher order of complexity. When the paper on *Pisum* was written Mendel apparently inclined to the view that with modifications his law might be found to include all the phenomena of hybridisation, but in the brief subsequent paper on *Hieracium* he clearly recognized the existence of cases of a different nature. Those who read that contribution will be interested to see that he lays down a principle which may be extended from hybridisation to heredity in general, that the laws of each new case must be determined by separate experiment.

As regards the Mendelian principles, which it is the chief aim of this introduction to present clearly before the reader, a professed student of variation will easily be able to fill in the outline now indicated, and to illustrate the various conceptions from phenomena already familiar. To do this is beyond the scope of this short sketch. But enough perhaps has now been said to show that by the application of those principles we are enabled to reach and deal in a comprehensive manner with phenomena of a fundamental nature, lying at the very root of all conceptions not merely of the physiology of reproduction and heredity, but even of the essential nature of living organisms; and I think that I used no extravagant words when, in introducing Mendel's work to the notice of readers of the Royal Horticultural Society's Journal, I ventured to declare that his experiments are worthy to rank with those which laid the foundation of the Atomic laws of Chemistry.

As some biographical particulars of this remarkable investigator will be welcome, I give the following brief notice, first published by Dr Correns on the authority of Dr von Schanz: Gregor Johann Mendel was born on July 22, 1822, at Heinzendorf bei Odrau, in Austrian Silesia. He was the son of well-to-do peasants. In 1843 he entered as a novice the "Königinkloster," an Augustinian foundation in Altbrünn. In 1847 he was ordained priest. From 1851 to 1853 he studied physics and natural science at Vienna. Thence he returned to his cloister and became a teacher in the Realschule at Brünn. Subsequently he was made Abbot, and died January 6, 1884. The experiments described in his papers were carried out in the garden of his Cloister. Besides the two papers on hybridisation, dealing respectively with *Pisum* and *Hieracium*, Mendel contributed two brief notes to the *Verh. Zool. bot. Verein*, Wien, on *Scopolia margaritalis* (1853, III., p. 116) and on *Bruchus pisi* (*ibid.* 1854, IV., p. 27). In these papers he speaks of himself as a pupil of Kollar.

Mendel published in the Brünn journal statistical observations of a meteorological character, but, so far as I am aware, no others relating to natural history. Dr Correns tells me that in the latter part of his life he engaged in the Ultramontane Controversy. He was for a time President of the Brünn Society*.

For the photograph of Mendel which forms the frontispiece to this work, I am indebted to the Very Rev. Dr Janeischek, the present Abbot of Brünn, who most kindly supplied it for this purpose.

So far as I have discovered there was, up to 1900, only one reference to Mendel's observations in scientific literature, namely that of Focke, *Pflanzenmischlinge*, 1881, p. 109,

* A few additional particulars are given in Tschermak's edition.

where it is simply stated that Mendel's numerous experiments on *Pisum* gave results similar to those obtained by Knight, but that he believed he had found constant numerical ratios among the types produced by hybridisation. In the same work a similar brief reference is made to the paper on *Hieracium*.

It may seem surprising that a work of such importance should so long have failed to find recognition and to become current in the world of science. It is true that the journal in which it appeared is scarce, but this circumstance has seldom long delayed general recognition. The cause is unquestionably to be found in that neglect of the experimental study of the problem of Species which supervened on the general acceptance of the Darwinian doctrines. The problem of Species, as Kölreuter, Gärtner, Naudin, Wichura, and the other hybridists of the middle of the nineteenth century conceived it, attracted thenceforth no workers. The question, it was imagined, had been answered and the debate ended. No one felt much interest in the matter. A host of other lines of work were suddenly opened up, and in 1865 the more original investigators naturally found those new methods of research more attractive than the tedious observations of the hybridisers, whose inquiries were supposed, moreover, to have led to no definite result.

Nevertheless the total neglect of such a discovery is not easy to account for. Those who are acquainted with the literature of this branch of inquiry will know that the French Academy offered a prize in 1861 to be awarded in 1862 on the subject "*Étudier les Hybrides végétaux au point de vue de leur fécondité et de la perpétuité de leurs caractères.*" This subject was doubtless chosen with reference to the experiments of Godron of Nancy and Naudin, then of Paris. Both these naturalists competed, and the accounts of the work of Godron on *Datura* and of Naudin on a number of species were published in the years 1864 and 1865 respectively. Both, especially the latter, are works of high consequence in the history of the science of heredity. In the latter paper Naudin clearly enuntiated what we shall henceforth know as the Mendelian conception of the dissociation of characters of cross-breds in the formation of the germ-cells, though apparently he never developed this conception.

In the year 1864, George Bentham, then President of the Linnean Society, took these treatises as the subject of his address to the Anniversary meeting on the 24 May, Naudin's work being known to him from an abstract, the full paper having not yet appeared. Referring to the hypothesis of dissociation which he fully described, he said that it appeared to be new and well supported, but required much more confirmation before it could be held as proven. (*J. Linn. Soc., Bot.*, VIII., *Proc.*, p. XIV.)

In 1865, the year of Mendel's communication to the Brünn Society, appeared Wichura's famous treatise on his experiments with *Salix* to which Mendel refers. There are passages in this memoir which come very near Mendel's principles, but it is evident from the plan of his experiments that Mendel had conceived the whole of his ideas before that date.

In 1868 appeared the first edition of Darwin's *Animals and Plants*, marking the very zenith of these studies, and thenceforth the decline in the experimental investigation of Evolution and the problem of Species has been steady. With the rediscovery and confirmation of Mendel's work by de Vries, Correns and Tschermak in 1900 a new era begins.

That Mendel's work, appearing as it did, at a moment

when several naturalists of the first rank were still occupied with these problems, should have passed wholly unnoticed, will always remain inexplicable, the more so as the Brünn Society exchanged its publications with most of the Academies of Europe, including both the Royal and Linnean Societies.

Naudin's views were well known to Darwin and are discussed in *Animals and Plants* (ed. 1885, II., p. 23); but, put forward as they were without full proof, they could not command universal credence. Gärtner, too, had adopted opposite views; and Wichura, working with cases of another order, had proved the fact that some hybrids breed true. Consequently it is not to be wondered at that Darwin was sceptical. Moreover, the Mendelian idea of the "hybrid-character," or heterozygous form, was unknown to him, a conception without which the hypothesis of dissociation of characters is quite imperfect.

Had Mendel's work come into the hands of Darwin, it is not too much to say that the history of the development of evolutionary philosophy would have been very different from that which we have witnessed.

EXPERIMENTS IN PLANT-HYBRIDISATION*.

By Gregor Mendel.

(*Read at the Meetings of the* 8*th February and* 8*th March*, 1865.)

Introductory Remarks.

Experience of artificial fertilisation, such as is effected with ornamental plants in order to obtain new variations in colour, has led to the experiments which will here be discussed. The striking regularity with which the same hybrid forms always reappeared whenever fertilisation took place between the same species induced further experiments to be undertaken, the object of which was to follow up the developments of the hybrids in their progeny.

To this object numerous careful observers, such as Kölreuter, Gärtner, Herbert, Lecoq, Wichura and others, have devoted a part of their lives with inexhaustible perseverance. Gärtner especially, in his work "Die Bastarderzeugung im Pflanzenreiche" (The Production of Hybrids in the Vegetable Kingdom), has recorded very valuable observations; and quite recently Wichura published the results of some profound investigations into the hybrids

* [This translation was made by the Royal Horticultural Society, and is reprinted with modifications and corrections, by permission. The original paper was published in the *Verh. naturf. Ver. in Brünn*, *Abhandlungen*, IV. 1865, which appeared in 1866.]

of the Willow. That, so far, no generally applicable law governing the formation and development of hybrids has been successfully formulated can hardly be wondered at by anyone who is acquainted with the extent of the task, and can appreciate the difficulties with which experiments of this class have to contend. A final decision can only be arrived at when we shall have before us the results of detailed experiments made on plants belonging to the most diverse orders.

Those who survey the work done in this department will arrive at the conviction that among all the numerous experiments made, not one has been carried out to such an extent and in such a way as to make it possible to determine the number of different forms under which the offspring of hybrids appear, or to arrange these forms with certainty according to their separate generations, or to definitely ascertain their statistical relations*.

It requires indeed some courage to undertake a labour of such far-reaching extent; it appears, however, to be the only right way by which we can finally reach the solution of a question the importance of which cannot be overestimated in connection with the history of the evolution of organic forms.

The paper now presented records the results of such a detailed experiment. This experiment was practically confined to a small plant group, and is now, after eight years' pursuit, concluded in all essentials. Whether the plan upon which the separate experiments were conducted and carried out was the best suited to attain the desired end is left to the friendly decision of the reader.

* [It is to the clear conception of these three primary necessities that the whole success of Mendel's work is due. So far as I know this conception was absolutely new in his day.]

Selection of the Experimental Plants.

The value and utility of any experiment are determined by the fitness of the material to the purpose for which it is used, and thus in the case before us it cannot be immaterial what plants are subjected to experiment and in what manner such experiments are conducted.

The selection of the plant group which shall serve for experiments of this kind must be made with all possible care if it be desired to avoid from the outset every risk of questionable results.

The experimental plants must necessarily—

1. Possess constant differentiating characters.
2. The hybrids of such plants must, during the flowering period, be protected from the influence of all foreign pollen, or be easily capable of such protection.

The hybrids and their offspring should suffer no marked disturbance in their fertility in the successive generations.

Accidental impregnation by foreign pollen, if it occurred during the experiments and were not recognized, would lead to entirely erroneous conclusions. Reduced fertility or entire sterility of certain forms, such as occurs in the offspring of many hybrids, would render the experiments very difficult or entirely frustrate them. In order to discover the relations in which the hybrid forms stand towards each other and also towards their progenitors it appears to be necessary that all members of the series developed in each successive generation should be, *without exception*, subjected to observation.

At the very outset special attention was devoted to the *Leguminosæ* on account of their peculiar floral structure.

Experiments which were made with several members of this family led to the result that the genus *Pisum* was found to possess the necessary conditions.

Some thoroughly distinct forms of this genus possess characters which are constant, and easily and certainly recognisable, and when their hybrids are mutually crossed they yield perfectly fertile progeny. Furthermore, a disturbance through foreign pollen cannot easily occur, since the fertilising organs are closely packed inside the keel and the anther bursts within the bud, so that the stigma becomes covered with pollen even before the flower opens. This circumstance is of especial importance. As additional advantages worth mentioning, there may be cited the easy culture of these plants in the open ground and in pots, and also their relatively short period of growth. Artificial fertilisation is certainly a somewhat elaborate process, but nearly always succeeds. For this purpose the bud is opened before it is perfectly developed, the keel is removed, and each stamen carefully extracted by means of forceps, after which the stigma can at once be dusted over with the foreign pollen.

In all, thirty-four more or less distinct varieties of Peas were obtained from several seedsmen and subjected to a two years' trial. In the case of one variety there were remarked, among a larger number of plants all alike, a few forms which were markedly different. These, however, did not vary in the following year, and agreed entirely with another variety obtained from the same seedsmen; the seeds were therefore doubtless merely accidentally mixed. All the other varieties yielded perfectly constant and similar offspring; at any rate, no essential difference was observed during two trial years. For fertilisation twenty-two of these were selected and cultivated during the whole period of the experiments. They remained constant without any exception.

Their systematic classification is difficult and uncertain. If we adopt the strictest definition of a species, according to which only those individuals belong to a species which under precisely the same circumstances display precisely similar characters, no two of these varieties could be referred to one species. According to the opinion of experts, however, the majority belong to the species *Pisum sativum*; while the rest are regarded and classed, some as sub-species of *P. sativum*, and some as independent species, such as *P. quadratum*, *P. saccharatum*, and *P. umbellatum*. The positions, however, which may be assigned to them in a classificatory system are quite immaterial for the purposes of the experiments in question. It has so far been found to be just as impossible to draw a sharp line between the hybrids of species and varieties as between species and varieties themselves.

Division and Arrangement of the Experiments.

If two plants which differ constantly in one or several characters be crossed, numerous experiments have demonstrated that the common characters are transmitted unchanged to the hybrids and their progeny; but each pair of differentiating characters, on the other hand, unite in the hybrid to form a new character, which in the progeny of the hybrid is usually variable. The object of the experiment was to observe these variations in the case of each pair of differentiating characters, and to deduce the law according to which they appear in the successive generations. The experiment resolves itself therefore into just as many separate experiments as there are constantly differentiating characters presented in the experimental plants.

The various forms of Peas selected for crossing showed differences in the length and colour of the stem; in the size and form of the leaves; in the position, colour, and size of the flowers; in the length of the flower stalk; in the colour, form, and size of the pods; in the form and size of the seeds; and in the colour of the seed-coats and the albumen [cotyledons]. Some of the characters noted do not permit of a sharp and certain separation, since the difference is of a "more or less" nature, which is often difficult to define. Such characters could not be utilised for the separate experiments; these could only be confined to characters which stand out clearly and definitely in the plants. Lastly, the result must show whether they, in their entirety, observe a regular behaviour in their hybrid unions, and whether from these facts any conclusion can be come to regarding those characters which possess a subordinate significance in the type

The characters which were selected for experiment relate:

1. To the *difference in the form of the ripe seeds.* These are either round or roundish, the wrinkling, when such occurs on the surface, being always only shallow; or they are irregularly angular and deeply wrinkled (*P. quadratum*).

2. To the *difference in the colour of the seed albumen* (endosperm)*. The albumen of the ripe seeds is either pale yellow, bright yellow and orange coloured, or it possesses a more or less intense green tint. This difference of colour is easily seen in the seeds as their coats are transparent.

* [Mendel uses the terms "albumen" and "endosperm" somewhat loosely to denote the cotyledons, containing food-material, within the seed.]

3. To the *difference in the colour of the seed-coat.* This is either white, with which character white flowers are constantly correlated; or it is grey, grey-brown, leather-brown, with or without violet spotting, in which case the colour of the standards is violet, that of the wings purple, and the stem in the axils of the leaves is of a reddish tint. The grey seed-coats become dark brown in boiling water.

4. To the *difference in the form of the ripe pods.* These are either simply inflated, never contracted in places; or they are deeply constricted between the seeds and more or less wrinkled (*P. saccharatum*).

5. To the *difference in the colour of the unripe pods.* They are either light to dark green, or vividly yellow, in which colouring the stalks, leaf-veins, and calyx participate*.

6. To the *difference in the position of the flowers.* They are either axial, that is, distributed along the main stem; or they are terminal, that is, bunched at the top of the stem and arranged almost in a false umbel; in this case the upper part of the stem is more or less widened in section (*P. umbellatum*)†.

7. To the *difference in the length of the stem.* The length of the stem‡ is very various in some forms; it is,

* One species possesses a beautifully brownish-red coloured pod, which when ripening turns to violet and blue. Trials with this character were only begun last year. [Of these further experiments it seems no account was published. Correns has since worked with such a variety.]

† [This is often called the Mummy Pea. It shows slight fasciation. The form I know has white standard and salmon-red wings.]

‡ [In my account of these experiments (*R.H.S. Journal*, vol. xxv. p. 54) I misunderstood this paragraph and took "axis" to mean the *floral* axis, instead of the main axis of the plant. The unit of measurement, being indicated in the original by a dash ('), I care-

however, a constant character for each, in so far that healthy plants, grown in the same soil, are only subject to unimportant variations in this character.

In experiments with this character, in order to be able to discriminate with certainty, the long axis of 6—7 ft. was always crossed with the short one of ¾ ft. to 1½ ft.

Each two of the differentiating characters enumerated above were united by cross-fertilisation. There were made for the

1st	trial	60	fertilisations	on	15 plants.
2nd	"	58	"	"	10 "
3rd	"	35	"	"	10 "
4th	"	40	"	"	10 "
5th	"	23	"	"	5 "
6th	"	34	"	"	10 "
7th	"	37	"	"	10 "

From a larger number of plants of the same variety only the most vigorous were chosen for fertilisation. Weakly plants always afford uncertain results, because even in the first generation of hybrids, and still more so in the subsequent ones, many of the offspring either entirely fail to flower or only form a few and inferior seeds.

Furthermore, in all the experiments reciprocal crossings were effected in such a way that each of the two varieties which in one set of fertilisations served as seed-bearers in the other set were used as pollen plants.

The plants were grown in garden beds, a few also in pots, and were maintained in their naturally upright position by means of sticks, branches of trees, and strings stretched between. For each experiment a number of pot plants were placed during the blooming period in a greenhouse, to serve as control plants for the main experiment

lessly took to have been an *inch*, but the translation here given is evidently correct.]

in the open as regards possible disturbance by insects. Among the insects* which visit Peas the beetle *Bruchus pisi* might be detrimental to the experiments should it appear in numbers. The female of this species is known to lay the eggs in the flower, and in so doing opens the keel; upon the tarsi of one specimen, which was caught in a flower, some pollen grains could clearly be seen under a lens. Mention must also be made of a circumstance which possibly might lead to the introduction of foreign pollen. It occurs, for instance, in some rare cases that certain parts of an otherwise quite normally developed flower wither, resulting in a partial exposure of the fertilising organs. A defective development of the keel has also been observed, owing to which the stigma and anthers remained partially uncovered†. It also sometimes happens that the pollen does not reach full perfection. In this event there occurs a gradual lengthening of the pistil during the blooming period, until the stigmatic tip protrudes at the point of the keel. This remarkable appearance has also been observed in hybrids of *Phaseolus* and *Lathyrus*.

The risk of false impregnation by foreign pollen is, however, a very slight one with *Pisum*, and is quite incapable of disturbing the general result. Among more than 10,000 plants which were carefully examined there were only a very few cases where an indubitable false impregnation had occurred. Since in the greenhouse such a case was never remarked, it may well be supposed that *Bruchus pisi*, and possibly also the described abnormalities in the floral structure, were to blame.

* [It is somewhat surprising that no mention is made of Thrips, which swarm in Pea flowers. I had come to the conclusion that this is a real source of error and I see Laxton held the same opinion.]

† [This also happens in Sweet Peas.]

The Forms of the Hybrids.*

Experiments which in previous years were made with ornamental plants have already afforded evidence that the hybrids, as a rule, are not exactly intermediate between the parental species. With some of the more striking characters, those, for instance, which relate to the form and size of the leaves, the pubescence of the several parts, &c., the intermediate, indeed, was nearly always to be seen; in other cases, however, one of the two parental characters was so preponderant that it was difficult, or quite impossible, to detect the other in the hybrid.

This is precisely the case with the Pea hybrids. In the case of each of the seven crosses the hybrid-character resembles† that of one of the parental forms so closely that the other either escapes observation completely or cannot be detected with certainty. This circumstance is of great importance in the determination and classification of the forms under which the offspring of the hybrids appear. Henceforth in this paper those characters which are transmitted entire, or almost unchanged in the hybridisation, and therefore in themselves constitute the characters of the hybrid, are termed the *dominant*, and those which become latent in the process *recessive.* The expression "recessive" has been chosen because the characters thereby designated withdraw or entirely disappear in the hybrids,

* [Mendel throughout speaks of his cross-bred Peas as "hybrids," a term which many restrict to the offspring of two distinct *species.* He, as he explains, held this to be only a question of degree.]

† [Note that Mendel, with true penetration, avoids speaking of the hybrid-character as "transmitted" by either parent, thus escaping the error pervading modern views of heredity.]

but nevertheless reappear unchanged in their progeny, as will be demonstrated later on.

It was furthermore shown by the whole of the experiments that it is perfectly immaterial whether the dominant character belong to the seed-bearer or to the pollen parent; the form of the hybrid remains identical in both cases. This interesting fact was also emphasised by Gärtner, with the remark that even the most practised expert is not in a position to determine in a hybrid which of the two parental species was the seed or the pollen plant*.

Of the differentiating characters which were used in the experiments the following are dominant:

1. The round or roundish form of the seed with or without shallow depressions.

2. The yellow colouring of the seed albumen [cotyledons].

3. The grey, grey-brown, or leather-brown colour of the seed-coat, in connection with violet-red blossoms and reddish spots in the leaf axils.

4. The simply inflated form of the pod.

5. The green colouring of the unripe pod in connection with the same colour in the stems, the leaf-veins and the calyx.

6. The distribution of the flowers along the stem.

7. The greater length of stem.

With regard to this last character it must be stated that the longer of the two parental stems is usually exceeded by the hybrid, which is possibly only attributable to the greater luxuriance which appears in all parts of plants when stems of very different length are crossed. Thus, for instance, in repeated experiments, stems of 1 ft. and 6 ft. in length yielded without exception hybrids which varied in length between 6 ft. and 7½ ft.

* [Gärtner, p. 223.]

The hybrid seeds in the experiments with seed-coat are often more spotted, and the spots sometimes coalesce into small bluish-violet patches. The spotting also frequently appears even when it is absent as a parental character.

The hybrid forms of the seed-shape and of the albumen are developed immediately after the artificial fertilisation by the mere influence of the foreign pollen. They can, therefore, be observed even in the first year of experiment, whilst all the other characters naturally only appear in the following year in such plants as have been raised from the crossed seed.

THE FIRST GENERATION [BRED] FROM THE HYBRIDS.

In this generation there reappear, together with the dominant characters, also the recessive ones with their full peculiarities, and this occurs in the definitely expressed average proportion of three to one, so that among each four plants of this generation three display the dominant character and one the recessive. This relates without exception to all the characters which were embraced in the experiments. The angular wrinkled form of the seed, the green colour of the albumen, the white colour of the seed-coats and the flowers, the constrictions of the pods, the yellow colour of the unripe pod, of the stalk of the calyx, and of the leaf venation, the umbel-like form of the inflorescence, and the dwarfed stem, all reappear in the numerical proportion given without any essential alteration. *Transitional forms were not observed in any experiment.*

Once the hybrids resulting from reciprocal crosses are fully formed, they present no appreciable difference in their subsequent development, and consequently the results [of the reciprocal crosses] can be reckoned together in each experiment. The relative numbers which were obtained for each pair of differentiating characters are as follows :

Expt. 1. Form of seed.—From 253 hybrids 7,324 seeds were obtained in the second trial year. Among them were 5,474 round or roundish ones and 1,850 angular wrinkled ones. Therefrom the ratio 2·96 to 1 is deduced.

Expt. 2. Colour of albumen.—258 plants yielded 8,023 seeds, 6,022 yellow, and 2,001 green ; their ratio, therefore, is as 3·01 to 1.

In these two experiments each pod yielded usually both kinds of seed. In well-developed pods which contained on the average six to nine seeds, it often occurred that all the seeds were round (Expt. 1) or all yellow (Expt. 2); on the other hand there were never observed more than five angular or five green ones in one pod. It appears to make no difference whether the pods are developed early or later in the hybrid or whether they spring from the main axis or from a lateral one. In some few plants only a few seeds developed in the first formed pods, and these possessed exclusively one of the two characters, but in the subsequently developed pods the normal proportions were maintained nevertheless.

As in separate pods, so did the distribution of the characters vary in separate plants. By way of illustration the first ten individuals from both series of experiments may serve*.

* [It is much to be regretted that Mendel does not give the complete series individually. No one who repeats such experiments should fail to record the *individual* numbers, which on seriation are sure to be full of interest.]

Plants.	Experiment 1. Form of Seed. Round.	Angular.	Experiment 2. Colour of Albumen. Yellow.	Green.
1	45	12	25	11
2	27	8	32	7
3	24	7	14	5
4	19	10	70	27
5	32	11	24	13
6	26	6	20	6
7	88	24	32	13
8	22	10	44	9
9	28	6	50	14
10	25	7	44	18

As extremes in the distribution of the two seed characters in one plant, there were observed in Expt. 1 an instance of 43 round and only 2 angular, and another of 14 round and 15 angular seeds. In Expt. 2 there was a case of 32 yellow and only 1 green seed, but also one of 20 yellow and 19 green.

These two experiments are important for the determination of the average ratios, because with a smaller number of experimental plants they show that very considerable fluctuations may occur. In counting the seeds, also, especially in Expt. 2, some care is requisite, since in some of the seeds of many plants the green colour of the albumen is less developed, and at first may be easily overlooked. The cause of the partial disappearance of the green colouring has no connection with the hybrid-character of the plants, as it likewise occurs in the parental variety. This peculiarity is also confined to the individual and is not inherited by the offspring. In luxuriant plants this appearance was frequently noted. Seeds which are damaged by insects during their development often vary in colour and form, but, with a little practice in sorting, errors are easily avoided. It is almost superfluous to mention that the pods must remain on the plants until they are thoroughly ripened and have become dried, since it is only then that the shape and colour of the seed are fully developed.

Expt. 3. Colour of the seed-coats.—Among 929 plants 705 bore violet-red flowers and grey-brown seed-coats; 224 had white flowers and white seed-coats, giving the proportion 3·15 to 1.

Expt. 4. Form of pods.—Of 1,181 plants 882 had them simply inflated, and in 299 they were constricted. Resulting ratio, 2·95 to 1.

Expt. 5. Colour of the unripe pods.—The number of trial plants was 580, of which 428 had green pods and 152 yellow ones. Consequently these stand in the ratio 2·82 to 1.

Expt. 6. Position of flowers.—Among 858 cases 651 blossoms were axial and 207 terminal. Ratio, 3·14 to 1.

Expt. 7. Length of stem.—Out of 1,064 plants, in 787 cases the stem was long, and in 277 short. Hence a mutual ratio of 2·84 to 1. In this experiment the dwarfed plants were carefully lifted and transferred to a special bed. This precaution was necessary, as otherwise they would have perished through being overgrown by their tall relatives. Even in their quite young state they can be easily picked out by their compact growth and thick dark-green foliage.

If now the results of the whole of the experiments be brought together, there is found, as between the number of forms with the dominant and recessive characters, an average ratio of 2·98 to 1, or 3 to 1.

The dominant character can have here a *double signification*—viz. that of a parental-character, or a hybrid-character*. In which of the two significations it appears in each separate case can only be determined by the following generation. As a parental character it must pass over unchanged to the whole of the offspring; as a hybrid-character, on the other hand, it must observe the same behaviour as in the first generation.

THE SECOND GENERATION [BRED] FROM THE HYBRIDS.

Those forms which in the first generation maintain the recessive character do not further vary in the second generation as regards this character; they remain constant in their offspring.

It is otherwise with those which possess the dominant character in the first generation [bred from the hybrids]. Of these *two*-thirds yield offspring which display the dominant and recessive characters in the proportion of 3 to 1, and thereby show exactly the same ratio as the hybrid forms, while only *one*-third remains with the dominant character constant.

The separate experiments yielded the following results:—

Expt. 1.—Among 565 plants which were raised from round seeds of the first generation, 193 yielded round seeds only, and remained therefore constant in this character; 372, however, gave both round and angular seeds, in the proportion of 3 to 1. The number of the hybrids, therefore, as compared with the constants is 1·93 to 1.

Expt. 2.—Of 519 plants which were raised from seeds whose albumen was of yellow colour in the first generation, 166 yielded exclusively yellow, while 353 yielded yellow

* [This paragraph presents the view of the hybrid-character as something incidental to the hybrid, and not "transmitted" to it—a true and fundamental conception here expressed probably for the first time.]

and green seeds in the proportion of 3 to 1. There resulted, therefore, a division into hybrid and constant forms in the proportion of 2·13 to 1.

For each separate trial in the following experiments 100 plants were selected which displayed the dominant character in the first generation, and in order to ascertain the significance of this, ten seeds of each were cultivated.

Expt. 3.—The offspring of 36 plants yielded exclusively grey-brown seed-coats, while of the offspring of 64 plants some had grey-brown and some had white.

Expt. 4.—The offspring of 29 plants had only simply inflated pods; of the offspring of 71, on the other hand, some had inflated and some constricted.

Expt. 5.—The offspring of 40 plants had only green pods; of the offspring of 60 plants some had green, some yellow ones.

Expt. 6.—The offspring of 33 plants had only axial flowers; of the offspring of 67, on the other hand, some had axial and some terminal flowers.

Expt. 7.—The offspring of 28 plants inherited the long axis, and those of 72 plants some the long and some the short axis.

In each of these experiments a certain number of the plants came constant with the dominant character. For the determination of the proportion in which the separation of the forms with the constantly persistent character results, the two first experiments are of especial importance, since in these a larger number of plants can be compared. The ratios 1·93 to 1 and 2·13 to 1 gave together almost exactly the average ratio of 2 to 1. The sixth experiment has a quite concordant result; in the others the ratio varies more or less, as was only to be expected in view of the smaller number of 100 trial plants. Experiment 5, which shows the greatest departure, was repeated, and then in lieu of the ratio of 60 and 40 that of 65 and 35 resulted. *The average ratio of* 2 *to* 1 *appears, therefore, as fixed with certainty.* It is therefore demonstrated that, of those forms which possess the dominant character in the first generation, in two-thirds the hybrid character is embodied, while one-third remains constant with the dominant character.

The ratio of 3 to 1, in accordance with which the distribution of the dominant and recessive characters results in the first generation, resolves itself therefore in all experiments into the ratio of 2 : 1 : 1 if the dominant character be differentiated according to its significance as a hybrid character or a parental one. Since the members of the first generation spring directly from the seed of the hybrids, *it is now clear that the hybrids form seeds having one or other of the two differentiating characters, and of these one-half develop again the hybrid form, while the other half yield plants which remain constant and receive the dominant or recessive characters [respectively] in equal numbers.*

The Subsequent Generations [Bred] from the Hybrids.

The proportions in which the descendants of the hybrids develop and split up in the first and second generations presumably hold good for all subsequent progeny. Experiments 1 and 2 have already been carried through six generations, 3 and 7 through five, and 4, 5, and 6 through four, these experiments being continued from the third generation with a small number of plants, and no departure from the rule has been perceptible. The offspring of the hybrids separated in each generation in the ratio of 2 : 1 : 1 into hybrids and constant forms.

If A be taken as denoting one of the two constant characters, for instance the dominant, a, the recessive, and Aa the hybrid form in which both are conjoined, the expression

$$A + 2Aa + a$$

shows the terms in the series for the progeny of the hybrids of two differentiating characters.

The observation made by Gärtner, Kölreuter, and others, that hybrids are inclined to revert to the parental forms, is also confirmed by the experiments described. It is seen that the number of the hybrids which arise from one fertilisation, as compared with the number of forms which become constant, and their progeny from generation to generation, is continually diminishing, but that nevertheless they could not entirely disappear. If an average equality of fertility in all plants in all generations be assumed, and if, furthermore, each hybrid forms seed of which one-half yields hybrids again, while the other half is constant to both characters in equal proportions, the ratio of numbers for the offspring in each generation is seen by the following summary, in which A and a denote again the two parental characters, and Aa the hybrid forms. For brevity's sake it may be assumed that each plant in each generation furnishes only 4 seeds.

				Ratios.				
Generation	A	Aa	a	A	:	Aa	:	a
1	1	2	1	1	:	2	:	1
2	6	4	6	3	:	2	:	3
3	28	8	28	7	:	2	:	7
4	120	16	120	15	:	2	:	15
5	496	32	496	31	:	2	:	31
n				$2^n - 1$	:	2	:	$2^n - 1$

In the tenth generation, for instance, $2^n - 1 = 1023$. There result, therefore, in each 2,048 plants which arise in this generation 1,023 with the constant dominant character, 1,023 with the recessive character, and only two hybrids.

The Offspring of Hybrids in which Several Differentiating Characters are Associated.

In the experiments above described plants were used which differed only in one essential character*. The next task consisted in ascertaining whether the law of development discovered in these applied to each pair of differentiating characters when several diverse characters are united in the hybrid by crossing. As regards the form of the hybrids in these cases, the experiments showed throughout that this invariably more nearly approaches to that one of the two parental plants which possesses the greater number of dominant characters. If, for instance, the seed plant has a short stem, terminal white flowers, and simply inflated pods; the pollen plant, on the other hand, a long stem, violet-red flowers distributed along the stem, and constricted pods; the hybrid resembles the seed parent only in the form of the pod; in the other characters it agrees with the pollen parent. Should one of the two parental types possess only dominant characters, then the hybrid is scarcely or not at all distinguishable from it.

* [This statement of Mendel's in the light of present knowledge is open to some misconception. Though his work makes it evident that such varieties may exist, it is very unlikely that Mendel could have had seven pairs of varieties such that the members of each pair differed from each other in *only* one considerable character (*wesentliches Merkmal*). The point is probably of little theoretical or practical consequence, but a rather heavy stress is thrown on "*wesentlich.*"]

Two experiments were made with a larger number of plants. In the first experiment the parental plants differed in the form of the seed and in the colour of the albumen; in the second in the form of the seed, in the colour of the albumen, and in the colour of the seed-coats. Experiments with seed characters give the result in the simplest and most certain way.

In order to facilitate study of the data in these experiments, the different characters of the seed plant will be indicated by *A*, *B*, *C*, those of the pollen plant by *a, b, c*, and the hybrid forms of the characters by *Aa*, *Bb*, and *Cc*.

Expt. 1.—*AB*, seed parents; *ab*, pollen parents;
A, form round; *a*, form angular;
B, albumen yellow. *b*, albumen green.

The fertilised seeds appeared round and yellow like those of the seed parents. The plants raised therefrom yielded seeds of four sorts, which frequently presented themselves in one pod. In all 556 seeds were yielded by 15 plants, and of these there were:—

315 round and yellow,
101 angular and yellow,
108 round and green,
32 angular and green.

All were sown the following year. Eleven of the round yellow seeds did not yield plants, and three plants did not form seeds. Among the rest:

38 had round yellow seeds . . . *AB*
65 round yellow and green seeds . . *ABb*
60 round yellow and angular yellow seeds . *AaB*
138 round yellow and green, angular yellow and green seeds . . . . *AaBb*.

From the angular yellow seeds 96 resulting plants bore seed, of which:

28 had only angular yellow seeds	*aB*
68 angular yellow and green seeds	*aBb*.

From 108 round green seeds 102 resulting plants fruited, of which:

35 had only round green seeds	*Ab*
67 round and angular green seeds	*Aab*.

The angular green seeds yielded 30 plants which bore seeds all of like character; they remained constant *ab*.

The offspring of the hybrids appeared therefore under nine different forms, some of them in very unequal numbers. When these are collected and co-ordinated we find:

38	plants	with	the	sign	*AB*
35	"	"	"		*Ab*
28	"	"	"		*aB*
30	"	"	"		*ab*
65	"	"	"		*ABb*
68	"	"	"		*aBb*
60	"	"	"		*AaB*
67	"	"	"		*Aab*
138	"	"	"		*AaBb*.

The whole of the forms may be classed into three essentially different groups. The first embraces those with the signs *AB*, *Ab*, *aB*, and *ab*: they possess only constant characters and do not vary again in the next generation. Each of these forms is represented on the average thirty-three times. The second group embraces the signs *ABb*, *aBb*, *AaB*, *Aab*: these are constant in one character and hybrid in another, and vary in the next generation only as regards the hybrid character. Each of these appears on an average sixty-five times. The form $AaBb$ occurs 138 times: it is hybrid in both characters, and behaves exactly as do the hybrids from which it is derived.

If the numbers in which the forms belonging to these classes appear be compared, the ratios of 1, 2, 4 are unmistakably evident. The numbers 32, 65, 138 present very fair approximations to the ratio numbers of 33, 66, 132.

The developmental series consists, therefore, of nine classes, of which four appear therein always once and are constant in both characters; the forms AB, ab, resemble the parental forms, the two others present combinations between the conjoined characters A, a, B, b, which combinations are likewise possibly constant. Four classes appear always twice, and are constant in one character and hybrid in the other. One class appears four times, and is hybrid in both characters. Consequently the offspring of the hybrids, if two kinds of differentiating characters are combined therein, are represented by the expression

$$AB + Ab + aB + ab + 2ABb + 2aBb + 2AaB + 2Aab + 4AaBb.$$

This expression is indisputably a combination series in which the two expressions for the characters A and a, B and b, are combined. We arrive at the full number of the classes of the series by the combination of the expressions:

$$A + 2\,Aa + a$$
$$B + 2\,Bb + b.$$

Second Expt.

ABC, seed parents;	abc, pollen parents;
A, form round;	a, form angular;
B, albumen yellow;	b, albumen green;
C, seed-coat grey-brown.	c, seed-coat white.

This experiment was made in precisely the same way as the previous one. Among all the experiments it demanded the most time and trouble. From 24 hybrids 687 seeds were obtained in all: these were all either spotted, grey-brown or grey-green, round or angular*. From these in the following year 639 plants fruited, and, as further investigation showed, there were among them:

8	plants	*ABC.*	22	plants	*ABCc.*	45	plants	*ABbCc.*
14	„	*ABc.*	17	„	*AbCc.*	36	„	*aBbCc.*
9	„	*AbC.*	25	„	*aBCc.*	38	„	*AaBCc.*
11	„	*Abc.*	20	„	*abCc.*	40	„	*AabCc.*
8	„	*aBC.*	15	„	*ABbC.*	49	„	*AabbC.*
10	„	*aBc.*	18	„	*ABbc.*	48	„	*AaBbc.*
10	„	*abC.*	19	„	*aBbC.*			
7	„	*abc.*	24	„	*aBbc.*			
			14	„	*AaBC.*	78	„	*AaBbCc.*
			18	„	*AaBc.*			
			20	„	*AabC.*			
			16	„	*Aabc.*			

The whole expression contains 27 terms. Of these 8 are constant in all characters, and each appears on the average 10 times; 12 are constant in two characters, and hybrid in the third; each appears on the average 19 times; 6 are constant in one character and hybrid in the other two; each appears on the average 43 times. One form appears 78 times and is hybrid in all of the characters. The ratios 10, 19, 43, 78 agree so closely with the ratios 10, 20, 40, 80, or 1, 2, 4, 8, that this last undoubtedly represents the true value.

The development of the hybrids when the original

* [Note that Mendel does not state the cotyledon-colour of the first crosses in this case; for as the coats were thick, it could not have been seen without opening or peeling the seeds.]

parents differ in three characters results therefore according to the following expression :

$ABC + ABc + AbC + Abc + aBC + aBc + abC + abc + 2\,ABCc + 2\,AbCc + 2\,aBCc + 2\,abCc + 2\,ABbC + 2\,ABbc + 2\,aBbC + 2\,aBbc + 2AaBC + 2\,AaBc + 2\,AabC + 2\,Aabc + 4\,ABbCc + 4\,aBbCc + 4\,AaBCc + 4\,AabCc + 4\,AaBbC + 4\,AaBbc + 8\,AaBbCc.$

Here also is involved a combination series in which the expressions for the characters A and a, B and b, C and c, are united. The expressions

$$A + 2\,Aa + a$$
$$B + 2\,Bb + b$$
$$C + 2\,Cc + c$$

give all the classes of the series. The constant combinations which occur therein agree with all combinations which are possible between the characters A, B, C, a, b, c; two thereof, ABC and abc, resemble the two original parental stocks.

In addition, further experiments were made with a smaller number of experimental plants in which the remaining characters by twos and threes were united as hybrids: all yielded approximately the same results. There is therefore no doubt that for the whole of the characters involved in the experiments the principle applies that *the offspring of the hybrids in which several essentially different characters are combined represent the terms of a series of combinations, in which the developmental series for each pair of differentiating characters are associated.* It is demonstrated at the same time that *the relation of each pair of different characters in hybrid union is independent of the other differences in the two original parental stocks.*

If n represent the number of the differentiating characters in the two original stocks, 3^n gives the number of terms of the combination series, 4^n the number of individuals which belong to the series, and 2^n the number of unions which remain constant. The series therefore embraces, if the original stocks differ in four characters, $3^4 = 81$ of classes, $4^4 = 256$ individuals, and $2^4 = 16$ constant forms; or, which is the same, among each 256 offspring of the hybrids there are 81 different combinations, 16 of which are constant.

All constant combinations which in Peas are possible by the combination of the said seven differentiating characters were actually obtained by repeated crossing. Their number is given by $2^7 = 128$. Thereby is simultaneously given the practical proof *that the constant characters which appear in the several varieties of a group of plants may be obtained in all the associations which are possible according to the [mathematical] laws of combination, by means of repeated artificial fertilisation.*

As regards the flowering time of the hybrids, the experiments are not yet concluded. It can, however, already be stated that the period stands almost exactly between those of the seed and pollen parents, and that the constitution of the hybrids with respect to this character probably happens in the same way as in the case of the other characters. The forms which are selected for experiments of this class must have a difference of at least twenty days from the middle flowering period of one to that of the other; furthermore, the seeds when sown must all be placed at the same depth in the earth, so that they may germinate simultaneously. Also, during the whole flowering period, the more important variations in temperature must be taken into account, and the partial hastening or delaying of the flowering which may result therefrom. It is clear that this experiment presents many difficulties to be overcome and necessitates great attention.

If we endeavour to collate in a brief form the results arrived at, we find that those differentiating characters which admit of easy and certain recognition in the experimental plants, all behave exactly alike in their hybrid associations. The offspring of the hybrids of each pair of differentiating characters are, one-half, hybrid again, while the other half are constant in equal proportions having the characters of the seed and pollen parents respectively. If several differentiating characters are combined by cross-fertilisation in a hybrid, the resulting offspring form the terms of a combination series in which the permutation series for each pair of differentiating characters are united.

The uniformity of behaviour shown by the whole of the characters submitted to experiment permits, and fully justifies, the acceptance of the principle that a similar relation exists in the other characters which appear less sharply defined in plants, and therefore could not be included in the separate experiments. An experiment with peduncles of different lengths gave on the whole a fairly satisfactory result, although the differentiation and serial arrangement of the forms could not be effected with that certainty which is indispensable for correct experiment.

The Reproductive Cells of Hybrids.

The results of the previously described experiments induced further experiments, the results of which appear fitted to afford some conclusions as regards the composition of the egg and pollen cells of hybrids. An important matter for consideration is afforded in *Pisum* by the circumstance that among the progeny of the hybrids constant forms appear, and that this occurs, too, in all combinations of the associated characters. So far as experience goes, we find it in every case confirmed that constant progeny can only be formed when the egg cells and the fertilising pollen are of like character, so that both are provided with the material for creating quite similar individuals, as is the case with the normal fertilisation of pure species*. We must therefore regard it as essential that exactly similar factors are at work also in the production of the constant forms in the hybrid plants. Since the various constant forms are produced in *one* plant, or even in *one* flower of a plant, the conclusion appears logical that in the ovaries of the hybrids there are formed as many sorts of egg cells, and in the anthers as many sorts of pollen cells, as there are possible constant combination forms, and that these egg and pollen cells agree in their internal composition with those of the separate forms.

In point of fact it is possible to demonstrate theoretically that this hypothesis would fully suffice to account for the development of the hybrids in the separate generations, if we might at the same time assume that the various kinds of egg and pollen cells were formed in the hybrids on the average in equal numbers†.

In order to bring these assumptions to an experimental proof, the following experiments were designed. Two forms which were constantly different in the form of the seed and the colour of the albumen were united by fertilisation.

If the differentiating characters are again indicated as *A*, *B*, *a*, *b*, we have:

AB, seed parent;	*ab*, pollen parent;
A, form round;	*a*, form angular;
B, albumen yellow.	*b*, albumen green.

* ["False hybridism" was of course unknown to Mendel.]

† [This and the preceding paragraph contain the essence of the Mendelian principles of heredity.]

The artificially fertilised seeds were sown together with several seeds of both original stocks, and the most vigorous examples were chosen for the reciprocal crossing. There were fertilised :

1. The hybrids with the pollen of *AB*.
2. The hybrids " " *ab*.
3. *AB* " " the hybrids.
4. *ab* " " the hybrids.

For each of these four experiments the whole of the flowers on three plants were fertilised. If the above theory be correct, there must be developed on the hybrids egg and pollen cells of the forms *AB*, *Ab*, *aB*, *ab*, and there would be combined :—

1. The egg cells *AB*, *Ab*, *aB*, *ab* with the pollen cells *AB*.

2. The egg cells *AB*, *Ab*, *aB*, *ab* with the pollen cells *ab*.

3. The egg cells *AB* with the pollen cells *AB*, *Ab*, *aB*, *ab*.

4. The egg cells *ab* with the pollen cells *AB*, *Ab*, *aB*, *ab*.

From each of these experiments there could then result only the following forms :—

1. *AB*, *ABb*, *AaB*, *AaBb*.
2. *AaBb*, *Aab*, *aBb*, *ab*.
3. *AB*, *ABb*, *AaB*, *AaBb*.
4. *AaBb*, *Aab*, *aBb*, *ab*.

If, furthermore, the several forms of the egg and pollen cells of the hybrids were produced on an average in equal numbers, then in each experiment the said four combinations should stand in the same ratio to each other. A perfect agreement in the numerical relations was, however, not to be expected, since in each fertilisation, even in normal cases, some egg cells remain undeveloped or subsequently die, and many even of the well-formed seeds fail to germinate when sown. The above assumption is also limited in so far that, while it demands the formation of an equal number of the various sorts of egg and pollen cells, it does not require that this should apply to each separate hybrid with mathematical exactness.

The first and second experiments had primarily the object of proving the composition of the hybrid egg cells, while the third and fourth experiments were to decide that of the pollen cells*. As is shown by the above demonstration the first and second experiments and the third and fourth experiments should produce precisely the same combinations, and even in the second year the result should be partially visible in the form and colour of the artificially fertilised seed. In the first and third experiments the dominant characters of form and colour, A and B, appear in each union, and are also partly constant and partly in hybrid union with the recessive characters a and b, for which reason they must impress their peculiarity upon the whole of the seeds. All seeds should therefore appear round and yellow, if the theory be justified. In the second and fourth experiments, on the other hand, one union is hybrid in form and in colour, and consequently the seeds are round and yellow; another is hybrid in form, but constant in the recessive character of colour, whence the seeds are round and green; the third is constant in the recessive character of form but hybrid in colour, consequently the seeds are

* [To prove, namely, that both were similarly differentiated, and not one or other only.]

angular and yellow; the fourth is constant in both recessive characters, so that the seeds are angular and green. In both these experiments there were consequently four sorts of seed to be expected—viz. round and yellow, round and green, angular and yellow, angular and green.

The crop fulfilled these expectations perfectly. There were obtained in the

1st Experiment, 98 exclusively round yellow seeds;
3rd ,, 94 ,, ,, ,, ,,

In the 2nd Experiment, 31 round and yellow, 26 round and green, 27 angular and yellow, 26 angular and green seeds.

In the 4th Experiment, 24 round and yellow, 25 round and green, 22 angular and yellow, 27 angular and green seeds.

A favourable result could now scarcely be doubted; the next generation must afford the final proof. From the seed sown there resulted for the first experiment 90 plants, and for the third 87 plants which fruited: these yielded for the—

1st Exp.	3rd Exp.		
20	25	round yellow seeds	*AB*
23	19	round yellow and green seeds . .	*ABb*
25	22	round and angular yellow seeds . .	*AaB*
22	21	round and angular green and yellow seeds	*AaBb*

In the second and fourth experiments the round and yellow seeds yielded plants with round and angular yellow and green seeds, *AaBb*.

From the round green seeds plants resulted with round and angular green seeds, *Aab*.

The angular yellow seeds gave plants with angular yellow and green seeds, *aBb*.

From the angular green seeds plants were raised which yielded again only angular and green seeds, *ab*.

Although in these two experiments likewise some seeds did not germinate, the figures arrived at already in the previous year were not affected thereby, since each kind of seed gave plants which, as regards their seed, were like each other and different from the others. There resulted therefore from the

2nd Exp.	4th Exp.		
31	24	plants of the form	*AaBb*
26	25	" "	*Aab*
27	22	" "	*aBb*
26	27	" "	*ab*

In all the experiments, therefore, there appeared all the forms which the proposed theory demands, and also in nearly equal numbers.

In a further experiment the characters of floral colour and length of stem were experimented upon, and selection so made that in the third year of the experiment each character ought to appear in half of all the plants if the above theory were correct. *A*, *B*, *a*, *b* serve again as indicating the various characters.

A, violet-red flowers. *a*, white flowers.
B, axis long. *b*, axis short.

The form *Ab* was fertilised with *ab*, which produced the hybrid *Aab*. Furthermore, *aB* was also fertilised with *ab*, whence the hybrid *aBb*. In the second year, for further fertilisation, the hybrid *Aab* was used as seed parent, and hybrid *aBb* as pollen parent.

Seed parent, *Aab*. Pollen parent, *aBb*.
Possible egg cells, *Abab*. Pollen cells, *aBab*.

From the fertilisation between the possible egg and pollen cells four combinations should result, viz.:—

$$AaBb + aBb + Aab + ab.$$

From this it is perceived that, according to the above theory, in the third year of the experiment out of all the plants

Half should have violet-red flowers (Aa),	Classes	1, 3
,, ,, ,, white flowers (a)	,,	2, 4
,, ,, ,, a long axis (Bb)	,,	1, 2
,, ,, ,, a short axis (b)	,,	3, 4

From 45 fertilisations of the second year 187 seeds resulted, of which only 166 reached the flowering stage in the third year. Among these the separate classes appeared in the numbers following:—

Class.	Colour of flower.	Stem.	
1	violet-red	long	47 times
2	white	long	40 ,,
3	violet-red	short	38 ,,
4	white	short	41 ,,

There consequently appeared—

The violet-red flower colour	(Aa)	in 85 plants.
,, white ,, ,,	(a)	in 81 ,,
,, long stem	(Bb)	in 87 ,,
,, short ,,	(b)	in 79 ,,

The theory adduced is therefore satisfactorily confirmed in this experiment also.

For the characters of form of pod, colour of pod, and position of flowers experiments were also made on a small scale, and results obtained in perfect agreement. All combinations which were possible through the union of the differentiating characters duly appeared, and in nearly equal numbers.

Experimentally, therefore, the theory is justified that *the pea hybrids form egg and pollen cells which, in their constitution, represent in equal numbers all constant forms which result from the combination of the characters when united in fertilisation.*

The difference of the forms among the progeny of the hybrids, as well as the respective ratios of the numbers in which they are observed, find a sufficient explanation in the principle above deduced. The simplest case is afforded by the developmental series of each pair of differentiating characters. This series is represented by the expression $A + 2Aa + a$, in which A and a signify the forms with constant differentiating characters, and Aa the hybrid form of both. It includes in three different classes four individuals. In the formation of these, pollen and egg cells of the form A and a take part on the average equally in the fertilisation; hence each form [occurs] twice, since four individuals are formed. There participate consequently in the fertilisation—

The pollen cells $A + A + a + a$
The egg cells $A + A + a + a$.

It remains, therefore, purely a matter of chance which of the two sorts of pollen will become united with each separate egg cell. According, however, to the law of probability, it will always happen, on the average of many cases, that each pollen form A and a will unite equally often with each egg cell form A and a, consequently one of the two pollen cells A in the fertilisation will meet with the egg cell A and the other with an egg cell a, and so likewise one pollen cell a will unite with an egg cell A, and the other with egg cell a.

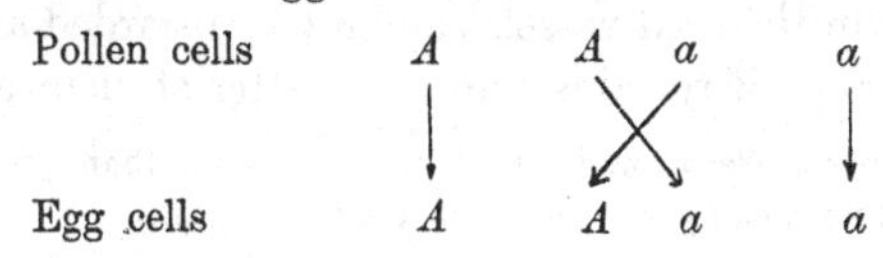

The result of the fertilisation may be made clear by putting the signs for the conjoined egg and pollen cells in the form of fractions, those for the pollen cells above and those for the egg cells below the line. We then have

$$\frac{A}{A}+\frac{A}{a}+\frac{a}{A}+\frac{a}{a}.$$

In the first and fourth term the egg and pollen cells are of like kind, consequently the product of their union must be constant, viz. A and a; in the second and third, on the other hand, there again results a union of the two differentiating characters of the stocks, consequently the forms resulting from these fertilisations are identical with those of the hybrid from which they sprang. *There occurs accordingly a repeated hybridisation.* This explains the striking fact that the hybrids are able to produce, besides the two parental forms, offspring which are like themselves; $\frac{A}{a}$ and $\frac{a}{A}$ both give the same union Aa, since, as already remarked above, it makes no difference in the result of fertilisation to which of the two characters the pollen or egg cells belong. We may write then—

$$\frac{A}{A}+\frac{A}{a}+\frac{a}{A}+\frac{a}{a}=A+2Aa+a.$$

This represents the average result of the self-fertilisation of the hybrids when two differentiating characters are united in them. In solitary flowers and in solitary plants, however, the ratios in which the forms of the series are produced may suffer not inconsiderable fluctuations*. Apart from the fact that the numbers in which both sorts of egg cells occur in the seed vessels can only be regarded as equal on the average, it remains purely a matter of chance which

* [Whether segregation by such units is more than purely fortuitous could probably be determined by seriation.]

of the two sorts of pollen may fertilise each separate egg cell. For this reason the separate values must necessarily be subject to fluctuations, and there are even extreme cases possible, as were described earlier in connection with the experiments on the form of the seed and the colour of the albumen. The true ratios of the numbers can only be ascertained by an average deduced from the sum of as many single values as possible; the greater the number the more are merely chance elements eliminated.

The developmental series for hybrids in which two kinds of differentiating characters are united contains among sixteen individuals nine different forms, viz.,

$AB + Ab + aB + ab + 2ABb + 2aBb + 2AaB + 2Aab + 4AaBb.$

Between the differentiating characters of the original stocks Aa and Bb four constant combinations are possible, and consequently the hybrids produce the corresponding four forms of egg and pollen cells AB, Ab, aB, ab, and each of these will on the average figure four times in the fertilisation, since sixteen individuals are included in the series. Therefore the participators in the fertilisation are—

Pollen cells $AB + AB + AB + AB + Ab + Ab + Ab + Ab +$
$aB + aB + aB + aB + ab + ab + ab + ab.$

Egg cells $AB + AB + AB + AB + Ab + Ab + Ab + Ab +$
$aB + aB + aB + aB + ab + ab + ab + ab.$

In the process of fertilisation each pollen form unites on an average equally often with each egg cell form, so that each of the four pollen cells AB unites once with one of the forms of egg cell AB, Ab, aB, ab. In precisely the same way the rest of the pollen cells of the forms Ab, aB, ab unite with all the other egg cells. We obtain therefore—

$$\frac{AB}{AB} + \frac{AB}{Ab} + \frac{AB}{aB} + \frac{AB}{ab} + \frac{Ab}{AB} + \frac{Ab}{Ab} + \frac{Ab}{aB} + \frac{Ab}{ab} +$$
$$\frac{aB}{AB} + \frac{aB}{Ab} + \frac{aB}{aB} + \frac{aB}{ab} + \frac{ab}{AB} + \frac{ab}{Ab} + \frac{ab}{aB} + \frac{ab}{ab},$$

or

$$AB + ABb + AaB + AaBb + ABb + Ab + AaBb + Aab + AaB + AaBb + aB + aBb + AaBb + Aab + aBb + ab = AB + Ab + aB + ab + 2ABb + 2aBb + 2AaB + 2Aab + 4AaBb^{*}.$$

In precisely similar fashion is the developmental series of hybrids exhibited when three kinds of differentiating characters are conjoined in them. The hybrids form eight various kinds of egg and pollen cells—*ABC*, *ABc*, *AbC*, *Abc*, *aBC*, *aBc*, *abC*, *abc*—and each pollen form unites itself again on the average once with each form of egg cell.

The law of combination of different characters which governs the development of the hybrids finds therefore its foundation and explanation in the principle enunciated, that the hybrids produce egg cells and pollen cells which in equal numbers represent all constant forms which result from the combinations of the characters brought together in fertilisation.

Experiments with Hybrids of other Species of Plants.

It must be the object of further experiments to ascertain whether the law of development discovered for *Pisum* applies also to the hybrids of other plants. To this end several experiments were recently commenced. Two minor experiments with species of *Phaseolus* have been completed, and may be here mentioned.

An experiment with *Phaseolus vulgaris* and *Phaseolus nanus* gave results in perfect agreement. *Ph. nanus* had together with the dwarf axis simply inflated green pods. *Ph. vulgaris* had, on the other hand, an axis 10 feet to

* [In the original the sign of equality (=) is here represented by +, evidently a misprint.]

12 feet high, and yellow coloured pods, constricted when ripe. The ratios of the numbers in which the different forms appeared in the separate generations were the same as with *Pisum*. Also the development of the constant combinations resulted according to the law of simple combination of characters, exactly as in the case of *Pisum*. There were obtained—

Constant combinations	Axis	Colour of the unripe pods.	Form of the ripe pods.
1	long	green	inflated
2	"	"	constricted
3	"	yellow	inflated
4	"	"	constricted
5	short	green	inflated
6	"	"	constricted
7	"	yellow	inflated
8	"	"	constricted

The green colour of the pod, the inflated forms, and the long axis were, as in *Pisum*, dominant characters.

Another experiment with two very different species of *Phaseolus* had only a partial result. *Phaseolus nanus*, L., served as seed parent, a perfectly constant species, with white flowers in short racemes and small white seeds in straight, inflated, smooth pods; as pollen parent was used *Ph. multiflorus*, W., with tall winding stem, purple-red flowers in very long racemes, rough, sickle-shaped crooked pods, and large seeds which bore black flecks and splashes on a peach-blood-red ground.

The hybrids had the greatest similarity to the pollen parent, but the flowers appeared less intensely coloured. Their fertility was very limited; from seventeen plants, which together developed many hundreds of flowers, only forty-nine seeds in all were obtained. These were of medium size, and were flecked and splashed similarly to those of *Ph. multiflorus*, while the ground colour was not materially different. The next year forty-four plants were raised from these seeds, of which only thirty-one reached the flowering stage. The characters of *Ph. nanus*, which had been altogether latent in the hybrids, reappeared in various combinations; their ratio, however, with relation to the dominant characters was necessarily very fluctuating owing to the small number of trial plants. With certain characters, as in those of the axis and the form of pod, it was, however, as in the case of *Pisum*, almost exactly 1 : 3.

Insignificant as the results of this experiment may be as regards the determination of the relative numbers in which the various forms appeared, it presents, on the other hand, the phenomenon of a remarkable change of colour in the flowers and seed of the hybrids. In *Pisum* it is known that the characters of the flower- and seed-colour present themselves unchanged in the first and further generations, and that the offspring of the hybrids display exclusively the one or the other of the characters of the original stocks*. It is otherwise in the experiment we are considering. The white flowers and the seed-colour of *Ph. nanus* appeared, it is true, at once in the first generation [*from* the hybrids] in one fairly fertile example, but the remaining thirty plants developed flower colours which were of various grades of purple-red to pale violet. The colouring of the seed-coat was no less varied than that of the flowers. No

* [This is the only passage where Mendel can be construed as asserting universal dominance for *Pisum*; and even here, having regard to the rest of the paper, it is clearly unfair to represent him as predicating more than he had seen in his own experiments. Moreover in flower and seed-coat colour (which is here meant), using his characters dominance must be almost universal, if not quite.]

plant could rank as fully fertile; many produced no fruit at all; others only yielded fruits from the flowers last produced, which did not ripen. From fifteen plants only were well-developed seeds obtained. The greatest disposition to infertility was seen in the forms with preponderantly red flowers, since out of sixteen of these only four yielded ripe seed. Three of these had a similar seed pattern to *Ph. multiflorus*, but with a more or less pale ground colour; the fourth plant yielded only one seed of plain brown tint. The forms with preponderantly violet coloured flowers had dark brown, black-brown, and quite black seeds.

The experiment was continued through two more generations under similar unfavourable circumstances, since even among the offspring of fairly fertile plants there were still some which were less fertile or even quite sterile. Other flower- and seed-colours than those cited did not subsequently present themselves. The forms which in the first generation [bred from the hybrids] contained one or more of the recessive characters remained, as regards these, constant without exception. Also of those plants which possessed violet flowers and brown or black seed, some did not vary again in these respects in the next generation; the majority, however, yielded, together with offspring exactly like themselves, some which displayed white flowers and white seed-coats. The red flowering plants remained so slightly fertile that nothing can be said with certainty as regards their further development.

Despite the many disturbing factors with which the observations had to contend, it is nevertheless seen by this experiment that the development of the hybrids, with regard to those characters which concern the form of the plants, follows the same laws as does *Pisum*. With regard to the colour characters, it certainly appears difficult to perceive a substantial agreement. Apart from the fact that from the union of a white and a purple-red colouring a whole series of colours results, from purple to pale violet and white, the circumstance is a striking one that among thirty-one flowering plants only one received the recessive character of the white colour, while in *Pisum* this occurs on the average in every fourth plant.

Even these enigmatical results, however, might probably be explained by the law governing *Pisum* if we might assume that the colour of the flowers and seeds of *Ph. multiflorus* is a combination of two or more entirely independent colours, which individually act like any other constant character in the plant. If the flower colour A were a combination of the individual characters $A_1 + A_2 + \ldots$ which produce the total impression of a purple colouration, then by fertilisation with the differentiating character, white colour, a, there would be produced the hybrid unions $A_1a + A_2a + \ldots$ and so would it be with the corresponding colouring of the seed-coats*. According to the above assumption, each of these hybrid colour unions would be independent, and would consequently develop quite independently from the others. It is then easily seen that from the combination of the separate developmental series

* [It appears to me clear that this expression is incorrectly given, and the argument regarding compound characters is consequently not legitimately developed. The original compound character should be represented as $A_1A_2A_3\ldots$ which when fertilised by a_1 gives $A_1A_2A_3\ldots a$ as the hybrid of the first generation. Mendel practically tells us these were all alike, and there is nothing to suggest that they were diverse. When on self-fertilisation, they break up, they will produce the gametes he specifies; but they may also produce A_1A_1 and A_2A_2, A_1A_2a, &c., thereby introducing terms of a nature different from any indicated by him. That this point is one of the highest significance, both practical and theoretical, is evident at once.]

a perfect colour-series must result. If, for instance, $A = A_1 + A_2$, then the hybrids A_1a and A_2a form the developmental series—

$$A_1 + 2A_1a + a$$
$$A_2 + 2A_2a + a.$$

The members of this series can enter into nine different combinations, and each of these denotes another colour*—

1 A_1A_2	2 A_1aA_2	1 A_2a
2 A_1A_2a	4 A_1aA_2a	2 A_2aa
1 A_1a	2 A_1aa	1 aa.

The figures prescribed for the separate combinations also indicate how many plants with the corresponding colouring belong to the series. Since the total is sixteen, the whole of the colours are on the average distributed over each sixteen plants, but, as the series itself indicates, in unequal proportions.

Should the colour development really happen in this way, we could offer an explanation of the case above described, viz. that the white flowers and seed-coat colour only appeared once among thirty-one plants of the first generation. This colouring appears only once in the series, and could therefore also only be developed once in the average in each sixteen, and with three colour characters only once even in sixty-four plants.

It must, however, not be forgotten that the explanation here attempted is based on a mere hypothesis, only supported by the very imperfect result of the experiment just described. It would, however, be well worth while to follow up the development of colour in hybrids by similar experi-

* [It seems very doubtful if the zygotes are correctly represented by the terms A_1aA_2a, A_2aa, A_1aa; for in the hybrids A_1a, &c. the allelomorphs A_1 and a, &c. should by hypothesis be separated in the gametes.]

B. 6

ments, since it is probable that in this way we might learn the significance of the extraordinary variety in the colouring of our ornamental flowers.

So far, little at present is known with certainty beyond the fact that the colour of the flowers in most ornamental plants is an extremely variable character. The opinion has often been expressed that the stability of the species is greatly disturbed or entirely upset by cultivation, and consequently there is an inclination to regard the development of cultivated forms as a matter of chance devoid of rules; the colouring of ornamental plants is indeed usually cited as an example of great instability. It is, however, not clear why the simple transference into garden soil should result in such a thorough and persistent revolution in the plant organism. No one will seriously maintain that in the open country the development of plants is ruled by other laws than in the garden bed. Here, as there, changes of type must take place if the conditions of life be altered, and the species possesses the capacity of fitting itself to its new environment. It is willingly granted that by cultivation the origination of new varieties is favoured, and that by man's labour many varieties are acquired which, under natural conditions, would be lost; but nothing justifies the assumption that the tendency to the formation of varieties is so extraordinarily increased that the species speedily lose all stability, and their offspring diverge into an endless series of extremely variable forms. Were the change in the conditions of vegetation the sole cause of variability we might expect that those cultivated plants which are grown for centuries under almost identical conditions would again attain constancy. That, as is well known, is not the case, since it is precisely under such circumstances that not only the most varied but also the most variable forms are found. It is only the *Leguminosæ*, like *Pisum, Phaseolus, Lens*, whose organs of fertilisation are protected by the keel, which constitute a noteworthy exception. Even here there have arisen numerous varieties during a cultural period of more than 1000 years; these maintain, however, under unchanging environments a stability as great as that of species growing wild.

It is more than probable that as regards the variability of cultivated plants there exists a factor which so far has received little attention. Various experiments force us to the conclusion that our cultivated plants, with few exceptions, are *members of various hybrid series*, whose further development in conformity with law is changed and hindered by frequent crossings *inter se*. The circumstance must not be overlooked that cultivated plants are mostly grown in great numbers and close together, affording the most favourable conditions for reciprocal fertilisation between the varieties present and the species itself. The probability of this is supported by the fact that among the great array of variable forms solitary examples are always found, which in one character or another remain constant, if only foreign influence be carefully excluded. These forms develop precisely as do those which are known to be members of the compound hybrid series. Also with the most susceptible of all characters, that of colour, it cannot escape the careful observer that in the separate forms the inclination to vary is displayed in very different degrees. Among plants which arise from *one* spontaneous fertilisation there are often some whose offspring vary widely in the constitution and arrangement of the colours, while others furnish forms of little deviation, and among a greater number solitary examples occur which transmit the colour of the flowers unchanged to their offspring. The cultivated species of *Dianthus*

6—2

afford an instructive example of this. A white-flowered example of *Dianthus caryophyllus*, which itself was derived from a white-flowered variety, was shut up during its blooming period in a greenhouse; the numerous seeds obtained therefrom yielded plants entirely white-flowered like itself. A similar result was obtained from a subspecies, with red flowers somewhat flushed with violet, and one with flowers white, striped with red. Many others, on the other hand, which were similarly protected, yielded progeny which were more or less variously coloured and marked.

Whoever studies the colouration which results in ornamental plants from similar fertilisation can hardly escape the conviction that here also the development follows a definite law which possibly finds its expression *in the combination of several independent colour characters.*

Concluding Remarks.

It can hardly fail to be of interest to compare the observations made regarding *Pisum* with the results arrived at by the two authorities in this branch of knowledge, Kölreuter and Gärtner, in their investigations. According to the opinion of both, the hybrids in outer appearance present either a form intermediate between the original species, or they closely resemble either the one or the other type, and sometimes can hardly be discriminated from it. From their seeds usually arise, if the fertilisation was effected by their own pollen, various forms which differ from the normal type. As a rule, the majority of individuals obtained by one fertilisation maintain the hybrid form, while some few others come more like the seed parent, and one or other individual approaches the pollen parent. This, however, is not the case with all hybrids without exception. With some the offspring have more nearly approached, some the one and some the other, original stock, or they all incline more to one or the other side; while with others *they remain perfectly like the hybrid* and continue constant in their offspring. The hybrids of varieties behave like hybrids of species, but they possess greater variability of form and a more pronounced tendency to revert to the original type.

With regard to the form of the hybrids and their development, as a rule an agreement with the observations made in *Pisum* is unmistakable. It is otherwise with the exceptional cases cited. Gärtner confesses even that the exact determination whether a form bears a greater resemblance to one or to the other of the two original species often involved great difficulty, so much depending upon the subjective point of view of the observer. Another circumstance could, however, contribute to render the results fluctuating and uncertain, despite the most careful observation and differentiation; for the experiments plants were mostly used which rank as good species and are differentiated by a large number of characters. In addition to the sharply defined characters, where it is a question of greater or less similarity, those characters must also be taken into account which are often difficult to define in words, but yet suffice, as every plant specialist knows, to give the forms a strange appearance. If it be accepted that the development of hybrids follows the law which is valid for *Pisum*, the series in each separate experiment must embrace very many forms, since the number of the components, as is known, increases with the number of the differentiating characters in *cubic ratio.* With a relatively small number of experimental-plants the result therefore could only be approximately right, and in single cases might fluctuate considerably. If, for instance, the two original stocks differ in seven characters, and 100 and 200 plants were raised from the seeds of their hybrids to determine the grade of relationship of the offspring, we can easily see how uncertain the decision must become, since for seven differentiating characters the combination series contains 16,384 individuals under 2187 various forms; now one and then another relationship could assert its predominance, just according as chance presented this or that form to the observer in a majority of cases.

If, furthermore, there appear among the differentiating characters at the same time dominant characters, which are transferred entire or nearly unchanged to the hybrids, then in the terms of the developmental series that one of the two original stocks which possesses the majority of dominant characters must always be predominant. In the experiment described relative to *Pisum*, in which three kinds of differentiating characters were concerned, all the dominant characters belonged to the seed parent. Although the terms of the series in their internal composition approach both original stock plants equally, in this experiment the type of the seed parent obtained so great a preponderance that out of each sixty-four plants of the first generation fifty-four exactly resembled it, or only differed in one character. It is seen how rash it may be under such circumstances to draw from the external resemblances of hybrids conclusions as to their internal nature.

Gärtner mentions that in those cases where the development was regular among the offspring of the hybrids the two original species were not reproduced, but only a few closely approximating individuals. With very extended developmental series it could not in fact be otherwise. For seven differentiating characters, for instance, among more than 16,000 individuals—offspring of the hybrids—

each of the two original species would occur only once. It is therefore hardly possible that these should appear at all among a small number of experimental plants; with some probability, however, we might reckon upon the appearance in the series of a few forms which approach them.

We meet with an *essential difference* in those hybrids which remain constant in their progeny and propagate themselves as truly as the pure species. According to Gärtner, to this class belong the *remarkably fertile hybrids Aquilegia atropurpurea canadensis, Lavatera pseudolbia thuringiaca, Geum urbano-rivale,* and some *Dianthus* hybrids; and, according to Wichura, the hybrids of the Willow species. For the history of the evolution of plants this circumstance is of special importance, since constant hybrids acquire the status of new species. The correctness of this is evidenced by most excellent observers, and cannot be doubted. Gärtner had opportunity to follow up *Dianthus Armeria deltoides* to the tenth generation, since it regularly propagated itself in the garden.

With *Pisum* it was shown by experiment that the hybrids form egg and pollen cells of *different* kinds, and that herein lies the reason of the variability of their offspring. In other hybrids, likewise, whose offspring behave similarly we may assume a like cause; for those, on the other hand, which remain constant the assumption appears justifiable that their fertilising cells are all alike and agree with the foundation-cell [fertilised ovum] of the hybrid. In the opinion of renowned physiologists, for the purpose of propagation one pollen cell and one egg cell unite in Phanerogams* into a single cell, which is capable by

* In *Pisum* it is placed beyond doubt that for the formation of the new embryo a perfect union of the elements of both fertilising cells must take place. How could we otherwise explain that among the

assimilation and formation of new cells to become an independent organism. This development follows a constant law, which is founded on the material composition and arrangement of the elements which meet in the cell in a vivifying union. If the reproductive cells be of the same kind and agree with the foundation cell [fertilised ovum] of the mother plant, then the development of the new individual will follow the same law which rules the mother plant. If it chance that an egg cell unites with a *dissimilar* pollen cell, we must then assume that between those elements of both cells, which determine the mutual differences, some sort of compromise is effected. The resulting compound cell becomes the foundation of the hybrid organism, the development of which necessarily follows a different scheme from that obtaining in each of the two original species. If the compromise be taken to be a complete one, in the sense, namely, that the hybrid embryo is formed from cells of like kind, in which the differences are *entirely and permanently accommodated* together, the further result follows that the hybrids, like any other stable plant species, remain true to themselves in their offspring. The reproductive cells which are formed in their seed

offspring of the hybrids both original types reappear in equal numbers and with all their peculiarities? If the influence of the egg cell upon the pollen cell were only external, if it fulfilled the *rôle* of a nurse only, then the result of each artificial fertilisation could be no other than that the developed hybrid should exactly resemble the pollen parent, or at any rate do so very closely. This the experiments so far have in no wise confirmed. An evident proof of the complete union of the contents of both cells is afforded by the experience gained on all sides that it is immaterial, as regards the form of the hybrid, which of the original species is the seed parent or which the pollen parent.

vessels and anthers are of one kind, and agree with the fundamental compound cell [fertilised ovum].

With regard to those hybrids whose progeny is *variable* we may perhaps assume that between the differentiating elements of the egg and pollen cells there also occurs a compromise, in so far that the formation of a cell as foundation of the hybrid becomes possible; but, nevertheless, the arrangement between the conflicting elements is only temporary and does not endure throughout the life of the hybrid plant. Since in the habit of the plant no changes are perceptible during the whole period of vegetation, we must further assume that it is only possible for the differentiating elements to liberate themselves from the enforced union when the fertilising cells are developed. In the formation of these cells all existing elements participate in an entirely free and equal arrangement, in which it is only the differentiating ones which mutually separate themselves. In this way the production would be rendered possible of as many sorts of egg and pollen cells as there are combinations possible of the formative elements.

The attribution attempted here of the essential difference in the development of hybrids to *a permanent or temporary union* of the differing cell elements can, of course, only claim the value of an hypothesis for which the lack of definite data offers a wide field. Some justification of the opinion expressed lies in the evidence afforded by *Pisum* that the behaviour of each pair of differentiating characters in hybrid union is independent of the other differences between the two original plants, and, further, that the hybrid produces just so many kinds of egg and pollen cells as there are possible constant combination forms. The differentiating characters of two plants can finally, however, only depend upon differences in the composition and grouping of the elements which exist in the foundation-cells [fertilised ova] of the same in vital interaction*.

Even the validity of the law formulated for *Pisum* requires still to be confirmed, and a repetition of the more important experiments is consequently much to be desired, that, for instance, relating to the composition of the hybrid fertilising cells. A differential [element] may easily escape the single observer†, which although at the outset may appear to be unimportant, may yet accumulate to such an extent that it must not be ignored in the total result. Whether the variable hybrids of other plant species observe an entire agreement must also be first decided experimentally. In the meantime we may assume that in material points a difference in principle can scarcely occur, since the unity in the developmental plan of organic life is beyond question.

In conclusion, the experiments carried out by Kölreuter, Gärtner, and others with respect to *the transformation of one species into another by artificial fertilisation* merit special mention. A special importance has been attached to these experiments, and Gärtner reckons them among "the most difficult of all in hybridisation."

If a species A is to be transformed into a species B, both must be united by fertilisation and the resulting hybrids then be fertilised with the pollen of B; then, out of the various offspring resulting, that form would be selected which stood in nearest relation to B and once more be fertilised with B pollen, and so continuously until finally a form is arrived at which is like B and constant in

* "*Welche in den Grundzellen derselben in lebendiger Wechselwirkung stehen.*"

† "*Dem einzelnen Beobachter kann leicht ein Differenziale entgehen.*"

its progeny. By this process the species *A* would change into the species *B*. Gärtner alone has effected thirty such experiments with plants of genera *Aquilegia*, *Dianthus*, *Geum*, *Lavatera*, *Lychnis*, *Malva*, *Nicotiana*, and *Œnothera*. The period of transformation was not alike for all species. While with some a triple fertilisation sufficed, with others this had to be repeated five or six times, and even in the same species fluctuations were observed in various experiments. Gärtner ascribes this difference to the circumstance that "the specific [*typische*] force by which a species, during reproduction, effects the change and transformation of the maternal type varies considerably in different plants, and that, consequently, the periods within which the one species is changed into the other must also vary, as also the number of generations, so that the transformation in some species is perfected in more, and in others in fewer generations." Further, the same observer remarks "that in these transformation experiments a good deal depends upon which type and which individual be chosen for further transformation."

If it may be assumed that in these experiments the constitution of the forms resulted in a similar way to that of *Pisum*, the entire process of transformation would find a fairly simple explanation. The hybrid forms as many kinds of egg cells as there are constant combinations possible of the characters conjoined therein, and one of these is always of the same kind as the fertilising pollen cells. Consequently there always exists the possibility with all such experiments that even from the second fertilisation there may result a constant form identical with that of the pollen parent. Whether this really be obtained depends in each separate case upon the number of the experimental plants, as well as upon the number of differentiating characters which are united by the fertilisation. Let us, for instance, assume that the plants selected for experiment differed in three characters, and the species *ABC* is to be transformed into the other species *abc* by repeated fertilisation with the pollen of the latter; the hybrids resulting from the first cross form eight different kinds of egg cells, viz.:

ABC, ABc, AbC, aBC, Abc, aBc, abC, abc.

These in the second year of experiment are united again with the pollen cells *abc*, and we obtain the series

$$AaBbCc + AaBbc + AabCc + aBbCc + Aabc + aBbc + abCc + abc.$$

Since the form *abc* occurs once in the series of eight components, it is consequently little likely that it would be missing among the experimental plants, even were these raised in a smaller number, and the transformation would be perfected already by a second fertilisation. If by chance it did not appear, then the fertilisation must be repeated with one of those forms nearest akin, *Aabc, aBbc, abCc.* It is perceived that such an experiment must extend the farther *the smaller the number of experimental plants and the larger the number of differentiating characters* in the two original species; and that, furthermore, in the same species there can easily occur a delay of one or even of two generations such as Gärtner observed. The transformation of widely divergent species could generally only be completed in five or six years of experiment, since the number of different egg cells which are formed in the hybrid increases in square ratio with the number of differentiating characters.

Gärtner found by repeated experiments that the respective period of transformation varies in many species, so that frequently a species *A* can be transformed into a species *B* a generation sooner than can species B into species A. He deduces therefrom that Kölreuter's opinion can hardly be maintained that "the two natures in hybrids are perfectly in equilibrium." It appears, however, that Kölreuter does not merit this criticism, but that Gärtner rather has overlooked a material point, to which he himself elsewhere draws attention, viz. that "it depends which individual is chosen for further transformation." Experiments which in this connection were carried out with two species of *Pisum* demonstrated that as regards the choice of the fittest individuals for the purpose of further fertilisation it may make a great difference which of two species is transformed into the other. The two experimental plants differed in five characters, while at the same time those of species A were all dominant and those of species B all recessive. For mutual transformation A was fertilised with pollen of B, and B with pollen of A, and this was repeated with both hybrids the following year. With the first experiment $\frac{B}{A}$ there were eighty-seven plants available in the third year of experiment for the selections of individuals for further crossing, and these were of the possible thirty-two forms; with the second experiment $\frac{A}{B}$ seventy-three plants resulted, which *agreed throughout perfectly in habit with the pollen parent*; in their internal composition, however, they must have been just as varied as the forms of the other experiment. A definite selection was consequently only possible with the first experiment; with the second some plants selected at random had to be excluded. Of the latter only a portion of the flowers were crossed with the A pollen, the others were left to fertilise themselves. Among each five plants which were selected in both experiments for fertilisation there agreed, as the following year's culture showed, with the pollen parent:—

1st Experiment.	2nd Experiment.	
2 plants	—	in all characters
3 ,,	—	,, 4 ,,
—	2 plants	,, 3 ,,
—	2 ,,	,, 2 ,,
—	1 plant	,, 1 character

In the first experiment, therefore, the transformation was completed; in the second, which was not continued further, two more fertilisations would probably have been required.

Although the case may not frequently occur that the dominant characters belong exclusively to one or the other of the original parent plants, it will always make a difference which of the two possesses the majority. If the pollen parent shows the majority, then the selection of forms for further crossing will afford a less degree of security than in the reverse case, which must imply a delay in the period of transformation, provided that the experiment is only considered as completed when a form is arrived at which not only exactly resembles the pollen plant in form, but also remains as constant in its progeny.

Gärtner, by the results of these transformation experiments, was led to oppose the opinion of those naturalists who dispute the stability of plant species and believe in a continuous evolution of vegetation. He perceives in the complete transformation of one species into another an indubitable proof that species are fixed within limits beyond which they cannot change. Although this opinion cannot be unconditionally accepted we find on the other hand in Gärtner's experiments a noteworthy confirmation of that supposition regarding variability of cultivated plants which has already been expressed.

Among the experimental species there were cultivated plants, such as *Aquilegia atropurpurea* and *canadensis*, *Dianthus caryophyllus*, *chinensis*, and *japonicus*, *Nicotiana rustica* and *paniculata*, and hybrids between these species lost none of their stability after four or five generations*.

* [The argument of these two last paragraphs appears to be that though the general mutability of natural species might be doubtful, yet among cultivated plants the transference of characters may be accomplished, and may occur by integral steps until one species is definitely "transformed" into the other.]

ON HIERACIUM-HYBRIDS OBTAINED BY ARTIFICIAL FERTILISATION

By G. Mendel.

(*Communicated to the Meeting* 9 *June*, 1869*.)

Although I have already undertaken many experiments in fertilisation between species of *Hieracium*, I have only succeeded in obtaining the following 6 hybrids, and only from one to three specimens of them.

H. Auricula ♀ × *H. aurantiacum* ♂
H. Auricula ♀ × *H. Pilosella* ♂
H. Auricula ♀ × *H. pratense* ♂
H. echioides† ♀ × *H. aurantiacum* ♂
H. prœaltum ♀ × *H. flagellare* Rchb. ♂
H. prœaltum ♀ × *H. aurantiacum* ♂

The difficulty of obtaining a larger number of hybrids is due to the minuteness of the flowers and their peculiar structure. On account of this circumstance it was seldom possible to remove the anthers from the flowers chosen for

* [Published in *Verh. naturf. Ver. Brünn, Abhandlungen*, VIII. 1869, p. 26, which appeared in 1870.]

† The plant used in this experiment is not exactly the typical *H. echioides*. It appears to belong to the series transitional to *H. prœaltum*, but approaches more nearly to *H. echioides* and for this reason was reckoned as belonging to the latter.

fertilisation without either letting pollen get on to the stigma or injuring the pistil so that it withered away. As is well known, the anthers are united to form a tube, which closely embraces the pistil. As soon as the flower opens, the stigma, already covered with pollen, protrudes. In order to prevent self-fertilisation the anther-tube must be taken out before the flower opens, and for this purpose the bud must be slit up with a fine needle. If this operation is attempted at a time when the pollen is mature, which is the case two or three days before the flower opens, it is seldom possible to prevent self-fertilisation; for with every care it is not easily possible to prevent a few pollen grains getting scattered and communicated to the stigma. No better result has been obtained hitherto by removing the anthers at an earlier stage of development. Before the approach of maturity the tender pistil and stigma are exceedingly sensitive to injury, and even if they are not actually injured, they generally wither and dry up after a little time if deprived of their protecting investments. I hope to obviate this last misfortune by placing the plants after the operation for two or three days in the damp atmosphere of a greenhouse. An experiment lately made with *H. Auricula* treated in this way gave a good result.

To indicate the object with which these fertilisation experiments were undertaken, I venture to make some preliminary remarks respecting the genus *Hieracium*. This genus possesses such an extraordinary profusion of distinct forms that no other genus of plants can compare with it. Some of these forms are distinguished by special peculiarities and may be taken as type-forms of species, while all the rest represent intermediate and transitional forms by which the type-forms are connected together. The difficulty in the separation and delimitation of these forms has demanded the close attention of the experts. Regarding no other genus has so much been written or have so many and such fierce controversies arisen, without as yet coming to a definite conclusion. It is obvious that no general understanding can be arrived at, so long as the value and significance of the intermediate and transitional forms is unknown.

Regarding the question whether and to what extent hybridisation plays a part in the production of this wealth of forms, we find very various and conflicting views held by leading botanists. While some of them maintain that this phenomenon has a far-reaching influence, others, for example, Fries, will have nothing to do with hybrids in *Hieracia*. Others take up an intermediate position; and while granting that hybrids are not rarely formed between the species in a wild state, still maintain that no great importance is to be attached to the fact, on the ground that they are only of short duration. The [suggested] causes of this are partly their restricted fertility or complete sterility; partly also the knowledge, obtained by experiment, that in hybrids self-fertilisation is always prevented if pollen of one of the parent-forms reaches the stigma. On these grounds it is regarded as inconceivable that *Hieracium* hybrids can constitute and maintain themselves as fully fertile and constant forms when growing near their progenitors.

The question of the origin of the numerous and constant intermediate forms has recently acquired no small interest since a famous *Hieracium* specialist has, in the spirit of the Darwinian teaching, defended the view that these forms are to be regarded as [arising] from the transmutation of lost or still existing species.

From the nature of the subject it is clear that without

an exact knowledge of the structure and fertility of the hybrids and the condition of their offspring through several generations no one can undertake to determine the possible influence exercised by hybridisation over the multiplicity of intermediate forms in *Hieracium*. The condition of the *Hieracium* hybrids in the range we are concerned with must necessarily be determined by experiments; for we do not possess a complete theory of hybridisation, and we may be led into erroneous conclusions if we take rules deduced from observation of certain other hybrids to be Laws of hybridisation, and try to apply them to *Hieracium* without further consideration. If by the experimental method we can obtain a sufficient insight into the phenomenon of hybridisation in *Hieracium*, then by the help of the experience which has been collected respecting the structural relations of the wild forms, a satisfactory judgment in regard to this question may become possible.

Thus we may express the object which was sought after in these experiments. I venture now to relate the very slight results which I have as yet obtained with reference to this object.

1. Respecting the structure of the hybrids, we have to record the striking phenomenon that the forms hitherto obtained by similar fertilisation are not identical. The hybrids *H. prœaltum* ♀ × *H. aurantiacum* ♂ and *H. Auricula* ♀ × *H. aurantiacum* ♂ are each represented by two, and *H. Auricula* ♀ × *H. pratense* ♂ by three individuals, while as to the remainder only one of each has been obtained.

If we compare the individual characters of the hybrids with the corresponding characters of the two parent types, we find that they sometimes present intermediate structures, but are sometimes so near to one of the parent characters that the [corresponding] character of the other has receded considerably or almost evades observation. So, for instance, we see in one of the two forms of *H. Auricula* ♀ × *H. aurantiacum* ♂ pure yellow disc-florets; only the petals of the marginal florets are on the outside tinged with red to a scarcely noticeable degree: in the other on the contrary the colour of these florets comes very near to *H. aurantiacum*, only in the centre of the disc the orange red passes into a deep golden-yellow. This difference is noteworthy, for the flower-colour in *Hieracium* has the value of a constant character. Other similar cases are to be found in the leaves, the peduncles, &c.

If the hybrids are compared with the parent types as regards the sum total of their characters, then the two forms of *H. prœaltum* ♀ × *H. aurantiacum* ♂ constitute approximately intermediate forms which do not agree in certain characters. On the contrary in *H. Auricula* ♀ × *H. aurantiacum* ♂ and in *H. Auricula* ♀ × *H. pratense* ♂ we see the forms widely divergent, so that one of them is nearer to the one and the other to the other parental type, while in the case of the last-named hybrid there is still a third which is almost precisely intermediate between them.

The conviction is then forced on us that we have here only single terms in an unknown series which may be formed by the direct action of the pollen of one species on the egg-cells of another.

2. With a single exception the hybrids in question form seeds capable of germination. *H. echioides* ♀ × *H. aurantiacum* ♂ may be described as fully fertile; *H. prœaltum* ♀ × *H. flagellare* ♂ as fertile; *H. prœaltum* ♀ × *H. aurantiacum* ♂ and *H. Auricula* ♀ × *H. pratense* ♂ as

partially fertile; *H. Auricula* ♀ × *H. Pilosella* ♂ as slightly fertile, and *H. Auricula* ♀ × *H. aurantiacum* ♂ as unfertile. Of the two forms of the last named hybrid, the red-flowered one was completely sterile, but from the yellow-flowered one a single well-formed seed was obtained. Moreover it must not pass unmentioned that among the seedlings of the partially fertile hybrid *H. prœaltum* ♀ × *H. aurantiacum* ♂ there was one plant which possessed full fertility.

[3.] As yet the offspring produced by self-fertilisation of the hybrids have not varied, but agree in their characters both with each other and with the hybrid plant from which they were derived.

From *H. prœaltum* ♀ × *H. flagellare* ♂ two generations have flowered; from *H. echioides* ♀ × *H. aurantiacum* ♂, *H. prœaltum* ♀ × *H. aurantiacum* ♂, *H. Auricula* ♀ × *H. Pilosella* ♂ one generation in each case has flowered.

4. The fact must be declared that in the case of the fully fertile hybrid *H. echioides* ♀ × *H. aurantiacum* ♂ the pollen of the parent types was not able to prevent self-fertilisation, though it was applied in great quantity to the stigmas protruding through the anther-tubes when the flowers opened.

From two flower-heads treated in this way seedlings were produced resembling this hybrid plant. A very similar experiment, carried out this summer with the partially fertile *H. prœaltum* ♀ × *H. aurantiacum* ♂ led to the conclusion that those flower-heads in which pollen of the parent type or of some other species had been applied to the stigmas, developed a notably larger number of seeds than those which had been left to self-fertilisation alone. The explanation of this result must only be sought in the circumstance that as a large part of the pollen-grains of the hybrid, examined microscopically, show a defective structure, a number of egg-cells capable of fertilisation do not become fertilised by their own pollen in the ordinary course of self-fertilisation.

It not rarely happens that in fully fertile species in the wild state the formation of the pollen fails, and in many anthers not a single good grain is developed. If in these cases seeds are nevertheless formed, such fertilisation must have been effected by foreign pollen. In this way hybrids may easily arise by reason of the fact that many forms of insects, notably the industrial Hymenoptera, visit the flowers of *Hieracia* with great zeal and are responsible for the pollen which easily sticks to their hairy bodies reaching the stigmas of neighbouring plants.

From the few facts that I am able to contribute it will be evident the work scarcely extends beyond its first inception. I must express some scruple in describing in this place an account of experiments just begun. But the conviction that the prosecution of the proposed experiments will demand a whole series of years, and the uncertainty whether it will be granted to me to bring the same to a conclusion have determined me to make the present communication. By the kindness of Dr Nägeli, the Munich Director, who was good enough to send me species which were wanting, especially from the Alps, I am in a position to include a larger number of forms in my experiments. I venture to hope even next year to be able to contribute something more by way of extension and confirmation of the present account.

If finally we compare the described result, still very uncertain, with those obtained by crosses made between forms of *Pisum*, which I had the honour of communicating in the year 1865, we find a very real distinction.

In *Pisum* the hybrids, obtained from the immediate crossing of two forms, have in all cases the same type, but their posterity, on the contrary, are variable and follow a definite law in their variations. In *Hieracium* according to the present experiments the exactly opposite phenomenon seems to be exhibited. Already in describing the *Pisum* experiments it was remarked that there are also hybrids whose posterity do not vary, and that, for example, according to Wichura the hybrids of *Salix* reproduce themselves like pure species. In *Hieracium* we may take it we have a similar case. Whether from this circumstance we may venture to draw the conclusion that the polymorphism of the genera *Salix* and *Hieracium* is connected with the special condition of their hybrids is still an open question, which may well be raised but not as yet answered.

A DEFENCE OF MENDEL'S PRINCIPLES OF HEREDITY.

"*The most fertile men of science have made blunders, and their consciousness of such slips has been retribution enough; it is only their more sterile critics who delight to dwell too often and too long on such mistakes.*" BIOMETRIKA, 1901.

INTRODUCTORY.

ON the rediscovery and confirmation of Mendel's Law by de Vries, Correns, and Tschermak two years ago, it became clear to many naturalists, as it certainly is to me, that we had found a principle which is destined to play a part in the Study of Evolution comparable only with the achievement of Darwin—that after the weary halt of forty years we have at last begun to march.

If we look back on the post-Darwinian period we recognize one notable effort to advance. This effort—fruitful as it proved, memorable as it must ever be—was that made by Galton when he enuntiated his Law of Ancestral Heredity, subsequently modified and restated by Karl Pearson. Formulated after long and laborious inquiry, this principle beyond question gives us an expression including and denoting many phenomena in which previously no regularity had been detected. But to practical naturalists it was evident from the first that there are great groups of facts which could not on any interpretation be brought within the scope of Galton's Law, and that by no emendation could that Law be extended to reach them. The existence of these phenomena pointed to a different physiological conception of heredity. Now it is precisely this conception that Mendel's Law enables us to form. Whether the Mendelian principle can be extended so as to include some apparently Galtonian cases is another question, respecting which we have as yet no facts to guide us, but we have certainly no warrant for declaring such an extension to be impossible.

Whatever answer the future may give to that question, it is clear from this moment that every case which obeys the Mendelian principle is removed finally and irretrievably from the operations of the Law of Ancestral Heredity.

At this juncture Professor Weldon intervenes as a professed exponent of Mendel's work. It is not perhaps to a devoted partisan of the Law of Ancestral Heredity that we should look for the most appreciative exposition of Mendel, but some bare measure of care and accuracy in representation is demanded no less in justice to fine work, than by the gravity of the issue.

Professor Weldon's article appears in the current number of *Biometrika*, Vol. I. Pt. II. which reached me on Saturday, Feb. 8. The paper opens with what purports to be a restatement of Mendel's experiments and results. In this "restatement" a large part of Mendel's experiments—perhaps the most significant—are not referred to at all. The perfect simplicity and precision of Mendel's own account are destroyed; with the result that the reader of Professor Weldon's paper, unfamiliar with Mendel's own memoir, can scarcely be blamed if he fail to learn the essence of the discovery. Of Mendel's conception of the hybrid as a distinct entity with characters proper to itself, apart from inheritance—the most novel thing in the whole paper—Professor Weldon gives no word. Upon this is poured an undigested mass of miscellaneous "facts" and statements from which the reader is asked to conclude, first, that a proposition attributed to Mendel regarding dominance of one character is not of "general"* application, and finally that "all work based on Mendel's method" is "vitiated" by a "fundamental mistake," namely "the neglect of ancestry†."

To find a parallel for such treatment of a great theme in biology we must go back to those writings of the orthodox which followed the appearance of the "Origin of Species."

On 17th December 1900 I delivered a Report to the Evolution Committee of the Royal Society on the experiments in Heredity undertaken by Miss E. R. Saunders and myself. This report has been offered to the Society for publication and will I understand shortly appear. In it we have attempted to show the extraordinary significance of Mendel's principle, to point out what in his results is essential and what subordinate, the ways in which the principle can be extended to apply to a diversity of more complex phenomena—of which some are incautiously cited

* The words "general" and "universal" appear to be used by Professor Weldon as interchangeable. Cp. Weldon, p. 235 and elsewhere, with Abstract given below.

† These words occur p. 252: "The fundamental mistake which vitiates all work based upon Mendel's method is the neglect of ancestry, and the attempt to regard the whole effect upon offspring produced by a particular parent, as due to the existence in the parent of particular structural characters, &c." As a matter of fact the view indicated in these last words is especially repugnant to the Mendelian principle, as will be seen.

by Professor Weldon as conflicting facts—and lastly to suggest a few simple terms without which (or some equivalents) the discussion of such phenomena is difficult. Though it is impossible here to give an outline of facts and reasoning there set out at length, I feel that his article needs an immediate reply. Professor Weldon is credited with exceptional familiarity with these topics, and his paper is likely to be accepted as a sufficient statement of the case. Its value will only be known to those who have either worked in these fields themselves or have been at the trouble of thoughtfully studying the original materials.

The nature of Professor Weldon's article may be most readily indicated if I quote the summary of it issued in a paper of abstracts sent out with Review copies of the Part. This paper was most courteously sent to me by an editor of *Biometrika* in order to call my attention to the article on Mendel, a subject in which he knew me to be interested. The abstract is as follows.

"Few subjects have excited so much interest in the last year or two as the laws of inheritance in hybrids. Professor W. F. R. Weldon describes the results obtained by Mendel by crossing races of Peas which differed in one or more of seven characters. From a study of the work of other observers, and from examination of the 'Telephone' group of hybrids, the conclusion is drawn that Mendel's results do not justify any general statement concerning inheritance in cross-bred Peas. A few striking cases of other cross-bred plants and animals are quoted to show that the results of crossing cannot, as Mendel and his followers suggest, be predicted from a knowledge of the characters of the two parents crossed without knowledge of the more remote ancestry."

Such is the judgment a fellow-student passes on this mind

"*Voyaging through strange seas of thought alone.*"

The only conclusion which most readers could draw from this abstract and indeed from the article it epitomizes, is that Mendel's discovery so far from being of paramount importance, rests on a basis which Professor Weldon has shown to be insecure, and that an error has come in through disregard of the law of Ancestral Heredity. On examining the paper it is perfectly true that Professor Weldon is careful nowhere directly to question Mendel's facts or his interpretation of them, for which indeed in some places he even expresses a mild enthusiasm, but there is no mistaking the general purpose of the paper. It must inevitably produce the impression that the importance of the work has been greatly exaggerated and that supporters of current views on Ancestry may reassure themselves. That this is Professor Weldon's own conclusion in the matter is obvious. After close study of his article it is evident to me that Professor Weldon's criticism is baseless and for the most part irrelevant, and I am strong in the conviction that the cause which will sustain damage from this debate is not that of Mendel.

I. The Mendelian Principle of Purity of Germ-Cells and the Laws of Heredity based on Ancestry.

Professor Weldon's article is entitled "Mendel's Laws of Alternative Inheritance in Peas." This title expresses the scope of Mendel's work and discovery none too precisely and even exposes him to distinct misconception.

To begin with, it says both too little and too much. Mendel did certainly determine Laws of Inheritance in peas—not precisely the laws Professor Weldon has been at the pains of drafting, but of that anon. Having done so, he knew what his discovery was worth. He saw, and rightly, that he had found a principle which *must* govern a wide area of phenomena. He entitles his paper therefore "*Versuche über Pflanzen-Hybriden*," or, Experiments in Plant-Hybridisation.

Nor did Mendel start at first with any particular intention respecting Peas. He tells us himself that he wanted to find the laws of inheritance in *hybrids*, which he suspected were definite, and that after casting about for a suitable subject, he found one in peas, for the reasons he sets out.

In another respect the question of title is much more important. By the introduction of the word "Alternative" the suggestion is made that the Mendelian principle applies peculiarly to cases of "alternative" inheritance. Mendel himself makes no such limitation in his earlier paper, though perhaps by rather remote implication in the second, to which the reader should have been referred. On the contrary, he wisely abstains from prejudicial consideration of unexplored phenomena.

To understand the significance of the word "alternative" as introduced by Professor Weldon we must go back a little in the history of these studies. In the year 1897 Galton formally announced the Law of Ancestral Heredity referred to in the *Introduction*, having previously "stated it briefly and with hesitation" in *Natural Inheritance*, p. 134. In 1898 Professor Pearson published his modification and generalisation of Galton's Law, introducing a correction of admitted theoretical importance, though it is not in question that the principle thus restated is fundamentally not very different from Galton's*. *It is an essential part of the Galton-Pearson Law of Ancestral Heredity that in calculating the probable structure of each descendant the structure of each several ancestor must be brought to account.*

Professor Weldon now tells us that these two papers of Galton and of Professor Pearson have "given us an expression for the effects of *blended* inheritance which seems likely to prove generally applicable, though the constants of the equations which express the relation between divergence from the mean in one generation, and that in another, may require modification in special cases. Our knowledge of *particulate* or mosaic inheritance, and of *alternative* inheritance, is however still rudimentary, and there is so much contradiction between the results obtained by different observers, that the evidence available is difficult to appreciate."

But Galton stated (p. 401) in 1897 that his statistical law of heredity "appears to be universally applicable to bi-sexual descent." Pearson in re-formulating the principle in 1898 made no reservation in regard to "alternative" inheritance. On the contrary he writes (p. 393) that "if Mr Galton's law can be firmly established, *it is a complete solution, at any rate to a first approximation, of the whole problem of heredity*," and again (p. 412) that "it is highly probable that it [this law] is the simple descriptive state-

* I greatly regret that I have not a precise understanding of the basis of the modification proposed by Pearson. His treatment is in algebraical form and beyond me. Nevertheless I have every confidence that the arguments are good and the conclusion sound. I trust it may not be impossible for him to provide the non-mathematical reader with a paraphrase of his memoir. The arithmetical differences between the original and the modified law are of course clear.

ment which brings into a single focus all the complex lines of hereditary influence. If Darwinian evolution be natural selection combined with *heredity*, then the single statement which embraces the whole field of heredity must prove almost as epoch-making as the law of gravitation to the astronomer*."

As I read there comes into my mind that other fine passage where Professor Pearson warns us

> "There is an insatiable desire in the human breast "to resume in some short formula, some brief "statement, the facts of human experience. It leads "the savage to 'account' for all natural phenomena "by deifying the wind and the stream and the tree. "It leads civilized man, on the other hand, to express "his emotional experience in works of art, and his "physical and mental experience in the formulae or "so-called laws of science†."

No naturalist who had read Galton's paper and had tried to apply it to the facts he knew could fail to see that here was a definite advance. We could all perceive phenomena that were in accord with it and there was no reasonable doubt that closer study would prove that accord to be close. It was indeed an occasion for enthusiasm, though no one acquainted with the facts of experimental breeding could consider the suggestion of universal application for an instant.

* I have searched Professor Pearson's paper in vain for any considerable reservation regarding or modification of this general statement. Professor Pearson enuntiates the law as "only correct on certain limiting hypotheses," but he declares that of these the most important is "the absence of reproductive selection, i.e. the negligible correlation of fertility with the inherited character, and the absence of sexual selection." The case of in-and-in breeding is also reserved.

† K. Pearson, *Grammar of Science*, 2nd ed. 1900, p. 36.

But two years have gone by, and in 1900 Pearson writes* that the values obtained from the Law of Ancestral Heredity

> "seem to fit the observed facts fairly well in the case of "*blended* inheritance. In other words we have a "certain amount of evidence in favour of the "conclusion: *That whenever the sexes are equipotent,* "*blend their characters and mate pangamously, all* "*characters will be inherited at the same rate,*"

or, again in other words, that the Law of Ancestral Heredity after the glorious launch in 1898 has been home for a complete refit. The top-hamper is cut down and the vessel altogether more manageable; indeed she looks trimmed for most weathers. Each of the qualifications now introduced wards off whole classes of dangers. Later on (pp. 487—8) Pearson recites a further list of cases regarded as exceptional. "All characters will be inherited at the same rate" might indeed almost be taken to cover the results in Mendelian cases, though the mode by which those results are arrived at is of course wholly different.

Clearly we cannot speak of the Law of Gravitation now. Our Tycho Brahe and our Kepler, with the yet more distant Newton, are appropriately named as yet to come†.

But the truth is that even in 1898 such a comparison was scarcely happy. Not to mention moderns, these high hopes had been finally disposed of by the work of the experimental breeders such as Kölreuter, Knight, Herbert, Gärtner, Wichura, Godron, Naudin, and many more. To have treated as non-existent the work of this group of naturalists, who alone have attempted to solve the problems

* *Grammar of Science,* 2nd ed. 1900, p. 480.

† *Phil. Trans.* 1900, vol. 195, A, p. 121.

of heredity and species—Evolution, as we should now say—by the only sound method—*experimental breeding*—to leave out of consideration almost the whole block of evidence collected in *Animals and Plants*—Darwin's finest legacy as I venture to declare—was unfortunate on the part of any exponent of Heredity, and in the writings of a professed naturalist would have been unpardonable. But even as modified in 1900 the Law of Ancestral Heredity is heavily over-sparred, and any experimental breeder could have increased Pearson's list of unconformable cases by as many again.

But to return to Professor Weldon. He now repeats that the Law of Ancestral Heredity seems likely to prove generally applicable to *blended* inheritance, but that the case of *alternative* inheritance is for the present reserved. We should feel more confidence in Professor Weldon's exposition if he had here reminded us that the special case which fitted Galton's Law so well that it emboldened him to announce that principle as apparently "universally applicable to bi-sexual descent" was one of *alternative* inheritance—namely the coat-colour of Basset-hounds. Such a fact is, to say the least, ominous. Pearson, in speaking (1900) of this famous case of Galton's, says that these phenomena of alternative inheritance must be treated separately (from those of blended inheritance)*, and for them he deduces a proposed "*law of reversion*," based of course on ancestry. He writes, "In both cases we may speak of a law of ancestral heredity, but the first predicts the probable character of the individual produced by a

* "If this be done, we shall, I venture to think, keep not only our minds, but our points for observation, clearer; and further, the failure of Mr Galton's statement in the one case will not in the least affect its validity in the other." Pearson (32), p. 143.

given ancestry, while the second tells us the percentages of the total offspring which on the average revert to each ancestral type*."

With the distinctions between the original Law of Ancestral Heredity, the modified form of the same law, and the Law of Reversion, important as all these considerations are, we are not at present concerned.

For the Mendelian principle of heredity asserts a proposition absolutely at variance with all the laws of ancestral heredity, however formulated. In those cases to which it applies strictly, this principle declares that the cross-breeding of parents *need* not diminish the purity of their germ-cells or consequently the purity of their offspring. When in such cases individuals bearing opposite characters, A and B, are crossed, the germ-cells of the resulting cross-bred, AB, are each to be bearers either of character A or of character B, not both.

Consequently when the cross-breds breed either together or with the pure forms, individuals will result of the forms AA, AB, BA, BB†. Of these the forms AA and BB, formed by the union of similar germs, are stated to be as pure as if they had had no cross in their pedigree, and henceforth their offspring will be no more likely to depart from the A type or the B type respectively, than those of any other originally pure specimens of these types.

Consequently in such examples it is *not* the fact that each ancestor must be brought to account as the Galton-Pearson Law asserts, and we are clearly dealing with a physiological phenomenon not contemplated by that Law at all.

* *Grammar of Science*, 1900, p. 494. See also Pearson, *Proc. Roy. Soc.* 1900, LXVI. pp. 142–3.

† On an average of cases, in equal numbers, as Mendel found.

Every case therefore which obeys the Mendelian principle is in direct contradiction to the proposition to which Professor Weldon's school is committed, and it is natural that he should be disposed to consider the Mendelian principle as applying especially to "alternative" inheritance, while the law of Galton and Pearson is to include the phenomenon of blended inheritance. The latter, he tells us, is "the most usual case," a view which, if supported by evidence, might not be without value.

It is difficult to blame those who on first acquaintance concluded Mendel's principle can have no strict application save to alternative inheritance. Whatever blame there is in this I share with Professor Weldon and those whom he follows. Mendel's own cases were almost all alternative; also the fact of dominance is very dazzling at first. But that was two years ago, and when one begins to see clearly again, it does not look so certain that the real essence of Mendel's discovery, the purity of germ-cells in respect of certain characters, may not apply also to some phenomena of blended inheritance. The analysis of this possibility would take us to too great length, but I commend to those who are more familiar with statistical method, the consideration of this question: whether dominance being absent, indefinite, or suppressed, the phenomena of heritages completely blended in the zygote, may not be produced by gametes presenting Mendelian purity of characters. A brief discussion of this possibility is given in the Introduction, p. 31.

Very careful inquiry would be needed before such a possibility could be negatived. For example, we know that the Laws based on Ancestry can apply to *alternative* inheritance; witness the case of the Basset-hounds. Here there is no simple Mendelian dominance; but are we sure there is no purity of germ-cells? The new conception goes a long way and it may well reach to such facts as these.

But for the present we will assume that Mendel's principle applies only to *certain phenomena of alternative inheritance*, which is as far as our warrant yet runs.

No close student of the recent history of evolutionary thought needs to be told what the attitude of Professor Weldon and his followers has been towards these same disquieting and unwelcome phenomena of alternative inheritance and discontinuity in variation. Holding at first each such fact for suspect, then treating them as rare and negligible occurrences, he and his followers have of late come slowly to accede to the facts of discontinuity a bare and grudging recognition in their scheme of evolution*.

Therefore on the announcement of that discovery which once and for all ratifies and consolidates the conception of discontinuous variation, and goes far to define that of alternative inheritance, giving a finite body to what before was vague and tentative, it is small wonder if Professor Weldon is disposed to criticism rather than to cordiality.

We have now seen what is the essence of Mendel's discovery based on a series of experiments of unequalled simplicity which Professor Weldon does not venture to dispute.

* Read in this connexion Pearson, K., *Grammar of Science*, 2nd ed. 1900, pp. 390—2.

Professor Weldon even now opens his essay with the statement—or perhaps reminiscence—that "it is perfectly possible and indeed probable that the difference between these forms of inheritance [blended, mosaic, and alternative] is only one of degree." This may be true; but reasoning favourable to this proposition could equally be used to prove the difference between mechanical mixture and chemical combination to be a difference of degree.

II. Mendel and the Critic's Version of him.

The "Law of Dominance."

I proceed to the question of dominance which Professor Weldon treats as a prime issue, almost to the virtual concealment of the great fact of gametic purity.

Cross-breds in general, AB and BA, named above, may present many appearances. They may all be indistinguishable from A, or from B; some may appear A's and some B's; they may be patchworks of both; they may be blends presenting one or many grades between the two; and lastly they *may have an appearance special to themselves* (*being in the latter case, as it often happens*, "*reversionary*"), a possibility which Professor Weldon does not stop to consider, though it is the clue that may unravel many of the facts which mystify him now.

Mendel's discovery became possible because he worked with regular cases of the first category, in which he was able to recognize that *one* of each of the pairs of characters he studied *did* thus prevail and *was* "dominant" in the cross-bred to the exclusion of the other character. This fact, which is still an accident of particular cases, Professor Weldon, following some of Mendel's interpreters, dignifies by the name of the "Law of Dominance," though he omits to warn his reader that Mendel states no "Law of Dominance" whatever. The whole question whether one or other character of the antagonistic pair is dominant though of great importance is logically a subordinate one. It depends on the specific nature of the varieties and individuals used, sometimes probably on the influence of external conditions and on other factors we cannot now discuss. There is as yet no universal law here perceived or declared.

Professor Weldon passes over the proof of the purity of the germ-cells lightly enough, but this proposition of dominance, suspecting its weakness, he puts prominently forward. Briefest equipment will suffice. Facing, as he supposes, some new pretender—some local Theudas—offering the last crazy prophecy,—any argument will do for such an one. An eager gathering in an unfamiliar literature, a scrutiny of samples, and he will prove to us with small difficulty that dominance of yellow over green, and round over wrinkled, is irregular even in peas after all; that in the sharpness of the discontinuity exhibited by the variations of peas there are many grades; that many of these grades co-exist in the same variety; that some varieties may perhaps be normally intermediate. All these propositions are supported by the production of a collection of evidence, the quality of which we shall hereafter consider. "Enough has been said," he writes (p. 240), "to show the grave discrepancy between the evidence afforded by Mendel's own experiments and that obtained by other observers, equally competent and trustworthy."

We are asked to believe that Professor Weldon has thus discovered "a fundamental mistake" vitiating all that work, the importance of which, he elsewhere tells us, he has "no wish to belittle."

III. The Facts in regard to Dominance of Characters in Peas.

Professor Weldon refers to no experiments of his own and presumably has made none. Had he done so he would have learnt many things about dominance in peas, whether of the yellow cotyledon-colour or of the round form, that might have pointed him to caution.

In the year 1900 Messrs Vilmorin-Andrieux & Co. were kind enough to send to the Cambridge Botanic Garden on my behalf a set of samples of the varieties of *Pisum* and *Phaseolus*, an exhibit of which had greatly interested me at the Paris Exhibition of that year. In the past summer I grew a number of these and made some preliminary cross-fertilizations among them (about 80 being available for these deductions) with a view to a future study of certain problems, Mendelian and others. In this work I had the benefit of the assistance of Miss Killby of Newnham College. Her cultivations and crosses were made independently of my own, but our results are almost identical. The experience showed me, what a naturalist would expect and practical men know already, that *a great deal turns on the variety used;* that some varieties are very sensitive to conditions while others maintain their type sturdily; that in using certain varieties Mendel's experience as to dominance is regularly fulfilled, while in the case of other varieties irregularities and even some contradictions occur. That the dominance of yellow cotyledon-colour over green, and the dominance of the smooth form over the wrinkled, is a *general* truth for *Pisum sativum* appears at once; that it is a universal truth I cannot believe any competent naturalist would imagine, still less assert. Mendel certainly never did.

When he speaks of the "law" or "laws" that he has established for *Pisum* he is referring to his own discovery of the purity of the germ-cells, that of the statistical distribution of characters among them, and the statistical grouping of the different germ-cells in fertilization, and not to the "Law of Dominance" which he never drafted and does not propound.

The issue will be clearer if I here state briefly what, as far as my experience goes, are the facts in regard to the characters *cotyledon-colour* and *seed-shapes* in peas. I have not opportunity for more than a passing consideration of the *seed-coats* of pure forms*; that is a maternal character, a fact I am not sure Professor Weldon fully appreciates. Though that may be incredible, it is evident from many passages that he has not, in quoting authorities, considered the consequences of this circumstance.

The normal characters: colour of cotyledons and seed-coats.

Culinary peas (*P. sativum*, omitting purple sorts) can primarily be classified on colour into two groups, yellow and green. In the green certain pigmentary matters persist in the ripe seed which disappear or are decomposed in the yellow as the seed ripens. But it may be observed

* The whole question as to seed-coat colour is most complex. Conditions of growth and ripening have a great effect on it. Mr Arthur Sutton has shown me samples of *Ne Plus Ultra* grown in England and abroad. This pea has yellow cotyledons with seed-coats either yellow or "blue." The foreign sample contained a much greater proportion of the former. He told me that generally speaking this is the case with samples ripened in a hot, dry climate.

Unquestionable Xenia appears occasionally, and will be spoken of later. Moreover to experiment with such a *plant*-character an extra generation has to be sown and cultivated. Consequently the evidence is meagre.

that the "green" class itself is treated as of two divisions, *green* and *blue*. In the seedsmen's lists the classification is made on the *external appearance* of the seed, without regard to whether the colour is due to the seed-coat, the cotyledons, or both. As a rule perhaps yellow coats contain yellow cotyledons, and green coats green cotyledons, though yellow cotyledons in green coats are common, e.g. *Gradus*, of which the cotyledons are yellow while the seed-coats are about as often green as yellow (or "white," as it is called technically). Those called "blue" consist mostly of seeds which have green cotyledons seen through transparent skins, or yellow cotyledons combined with green skins. The skins may be roughly classified into thin and transparent, or thick and generally at some stage pigmented. In numerous varieties the colour of the cotyledon is wholly yellow, or wholly green. Next there are many varieties which are constant in habit and other properties but have seeds belonging to these two colour categories in various proportions. How far these proportions are known to be constant I cannot ascertain.

Of such varieties showing mixture of *cotyledon*-colours nearly all can be described as dimorphic in colour. For example in Sutton's *Nonpareil Marrowfat* the cotyledons are almost always *either* yellow *or* green, with some piebalds, and the colours of the seed-coats are scarcely less distinctly dimorphic. In some varieties which exist in both colours intermediates are so common that one cannot assert any regular dimorphism*.

* Knowing my interest in this subject Professor Weldon was so good as to forward to me a series of his peas arranged to form a scale of colours and shapes, as represented in his Plate I. I have no doubt that the use of such colour-scales will much facilitate future study of these problems.

There are some varieties which have cotyledons green and intermediate shading to greenish yellow, like *Stratagem* quoted by Professor Weldon. Others have yellow and intermediate shading to yellowish green, such as McLean's *Best of all**. I am quite disposed to think there may be truly monomorphic varieties with cotyledons permanently of intermediate colour only, but so far I have not seen one†. The variety with greatest *irregularity* (apart from regular dimorphism) in cotyledon-colour I have seen is a sample of "*mange-tout à rames, à grain vert,*" but it was a good deal injured by weevils (*Bruchus*), which always cause irregularity or change of colour.

Lastly in some varieties there are many piebalds or mosaics.

From what has been said it will be evident that the description of a pea in an old book as having been green, blue, white, and so forth, unless the cotyledon-colour is distinguished from seed-coat colour, needs careful consideration before inferences are drawn from it.

Shape.

In regard to shape, if we keep to ordinary shelling peas, the facts are somewhat similar, but as shape is probably more sensitive to conditions than cotyledon-colour (not than *seed-coat* colour) there are irregularities to be perhaps ascribed to this cause. Broadly, however, there are two main divisions, round and wrinkled. It is unquestioned that between these two types every intermediate occurs.

* I notice that Vilmorin in the well-known *Plantes Potagères*, 1883, classifies the intermediate-coloured peas with the *green*.

† Similarly though *tall* and *dwarf* are Mendelian characters, peas occur of all heights and are usually classified as tall, half-dwarfs, and dwarfs.

Here again a vast number of varieties can be at once classified into round and wrinkled (the classification commonly used), others are intermediate normally. Here also I suspect some fairly clear sub-divisions might be made in the wrinkled group and in the round group too, but I would not assert this as a fact.

I cannot ascertain from botanists what is the nature of the difference between round and wrinkled peas, though no doubt it will be easily discovered. In maize the round seeds contain much unconverted starch, while in the wrinkled or sugar-maize this seems to be converted in great measure as the seed ripens; with the result that, on drying, the walls collapse. In such seeds we may perhaps suppose that the process of conversion, which in round seeds takes place on germination, is begun earlier, and perhaps the variation essentially consists in the premature appearance of the converting ferment. It would be most rash to suggest that such a process may be operating in the pea, for the phenomenon may have many causes; but however that may be, there is evidently a difference of such a nature that when the water dries out of the seed on ripening, its walls collapse*; and this collapse may occur in varying degrees.

* Wrinkling must of course be distinguished further from the squaring due to the peas pressing against each other in the pod.

In connexion with these considerations I may mention that Vilmorin makes the interesting statement that most peas retain their vitality three years, dying as a rule rapidly after that time is passed, though occasionally seeds seven or eight years old are alive; but that *wrinkled* peas germinate as a rule less well than round, and do not retain their vitality so long as the round. Vilmorin-Andrieux, *Plantes Potagères*, 1883, p. 423. Similar statements regarding the behaviour of wrinkled peas in India are made by Firminger, *Gardening for India*, 3rd ed. 1874, p. 146.

In respect of *shape* the seeds of a variety otherwise stable are as a rule fairly uniform, the co-existence of both shapes and of intermediates between them in the same variety is not infrequent. As Professor Weldon has said, *Telephone* is a good example of an extreme case of mixture of both colours and shapes. *William I.* is another. It may be mentioned that regular dimorphism in respect of shape is not so common as dimorphism in respect of colour. Of great numbers of varieties seen at Messrs Suttons' I saw none so distinctly dimorphic in shape as *William I.* which nevertheless contains all grades commonly.

So far I have spoken of the shapes of ordinary English culinary peas. But if we extend our observations to the shapes of *large-seeded* peas, which occur for the most part among the sugar-peas (*mange-touts*), of the "grey" peas with coloured flowers, etc., there are fresh complications to be considered.

Professor Weldon does not wholly avoid these (as Mendel did in regard to shape) and we will follow him through his difficulties hereafter. For the present let me say that the classes *round* and *wrinkled* are not readily applicable to those other varieties and are not so applied either by Mendel or other practical writers on these subjects. To use the terms indicated in the Introduction, *seed-shape* depends on more than one pair of allelomorphs—possibly on several.

Stability and Variability.

Generally speaking peas which when seen in bulk are monomorphic in colour and shape, will give fairly true and uniform offspring (but such strict monomorphism is rather exceptional). Instances to the contrary occur, and in my own brief experience I have seen some. In a row of *Fill-*

basket grown from selected seed there were two plants of different habit, seed-shape, etc. Each bore pods with seeds few though large and round. Again *Blue Peter* (blue and round) and *Laxton's Alpha* (blue and wrinkled), grown in my garden and left to nature uncovered, have each given a considerable proportion of seeds with *yellow* cotyledons, about 20 % in the case of *Laxton's Alpha.* The distribution of these on the plants I cannot state. The plants bearing them in each case sprang from green-cotyledoned seeds taken from samples containing presumably unselected green seeds only. A part of this exceptional result may be due to crossing, but heterogeneity of conditions* especially in or after ripening is a more likely cause, hypotheses I hope to investigate next season. Hitherto I had supposed the crossing, if any, to be done by *Bruchus* or Thrips, but Tschermak also suspects *Megachile,* the leaf-cutter bee, which abounds in my garden.

Whatever the cause, these irregularities may undoubtedly occur; and if they be proved to be largely independent of crossing and conditions, this will in nowise vitiate the truth of the Mendelian principle. For in that case it may simply be variability. Such true variation, or sporting, in the pea is referred to by many observers. Upon this subject I have received most valuable facts from Mr Arthur Sutton, who has very kindly interested himself in these inquiries.

* Cotyledon-colour is not nearly so sensitive to ordinary changes in conditions as coat-colour, provided the coat be uninjured. But even in monomorphic *green* varieties, a seed which for any cause has burst on ripening, has the exposed parts of its cotyledons *yellow.* The same may be the case in seeds of green varieties injured by *Bruchus* or birds. These facts make one hesitate before denying the effects of conditions on the cotyledon-colour even of uninjured seeds, and the variation described above may have been simply weathering. The seeds were gathered very late and many were burst in *Laxton's Alpha.* I do not yet know they are alive.

He tells me that several highly bred varieties, selected with every possible care, commonly throw a small but constant proportion of poor and almost vetch-like plants, with short pods and small round seeds, which are hoed out by experienced men each year before ripening. Other high-class varieties always, wherever grown, and when far from other sorts, produce a small percentage of some one or more definite "sports." Of these peculiar sports he has sent me a collection of twelve, taken from as many standard varieties, each "sport" being represented by eight seeds, which though quite distinct from the type agree with each other in almost all cases.

In two cases, he tells me, these seed-sports sown separately have been found to give plants identical with the standard type and must therefore be regarded as sports in *seed characters* only; in other cases change of plant-type is associated with the change of seed-type.

In most standard varieties these definite sports are not very common, but in a few they are common enough to require continual removal by selection*.

I hope before long to be able to give statistical details

* It is interesting to see that in at least one case the same—or practically the same—variety has been independently produced by different raisers, as we now perceive, by the fortuitous combination of similar allelomorphs. *Sutton's Ringleader* and *Carter's First Crop* (and two others) are cases in point, and it is peculiarly instructive to see that in the discussion of these varieties when they were new, one of the points indicating their identity was taken to be the fact that they produced *the same* "*rogues.*" See *Gard. Chron.* 1865, pp. 482 and 603; 1866, p. 221; 1867, pp. 546 and 712.

Rimpau quotes Blomeyer (*Kultur der Landw. Nutzpflanzen*, Leipzig, 1889, pp. 357 and 380) to the effect that *purple*-flowered plants with *wrinkled* seeds may spring as direct sports from peas with *white* flowers and *round* seeds. I have not seen a copy of Blomeyer's work. Probably this "wrinkling" was "indentation."

and experiments relating to this extraordinarily interesting subject. As de Vries writes in his fine work *Die Mutationstheorie* (I. p. 580), "a study of the seed-differences of inconstant, or as they are called, 'still' unfixed varieties, is a perfect treasure-house of new discoveries."

Let us consider briefly the possible significance of these facts in the light of Mendelian teaching. First, then, it is clear that as regards most of such cases the hypothesis is not excluded that these recurring sports may be due to the fortuitous concurrence of certain scarcer hypallelomorphs, which may either have been free in the original parent varieties from which the modern standard forms were raised, or may have been freed in the crossing to which the latter owe their origin (see p. 28). This possibility raises the question whether, if we could make "*pure* cultures" of the gametes, any variations of this nature would ever occur. This may be regarded as an unwarrantable speculation, but it is not wholly unamenable to the test of experiments.

But variability, in the sense of division of gonads into heterogeneous gametes, may surely be due to causes other than crossing. This we cannot doubt. Cross-fertilization of the zygote producing those gametes is *one* of the causes of such heterogeneity among them. We cannot suppose it to be the sole cause of this phenomenon.

When Mendel asserts the purity of the germ-cells of cross-breds he cannot be understood to mean that they are *more pure* than those of the original parental races. These must have varied in the past. The wrinkled seed arose from the round, the green from the yellow (or *vice versâ*, if preferred), and probably numerous intermediate forms from both.

The variations, or as I provisionally conceive it, that differentiant division among the gametes of which variation (neglecting environment) is the visible expression, has arisen and can arise at one or more points of time, and we have no difficulty in believing it to occur now. In many cases we have clear evidence that it does. Crossing,—dare we call it asymmetrical fertilization?—is *one* of the causes of the production of heterogeneous gametes—the result of divisions qualitatively differentiant and perhaps asymmetrical*.

There are other causes and we have to find them. Some years ago I wrote that consideration of the causes of variation was in my judgment premature†. Now that through Mendel's work we are clearing our minds as to the fundamental nature of "gametic" variation, the time is approaching when an investigation of such causes may be not unfruitful.

Of *variation* as distinct from *transmission* why does Professor Weldon take no heed? He writes (p. 244):

> "If Mendel's statements were universally valid, even among Peas, the characters of the seeds in the numerous hybrid races now existing should fall into one or other of a few definite categories, which should not be connected by intermediate forms."

Now, as I have already pointed out, Mendel made no pretence of universal statement: but had he done so, the conclusion, which Professor Weldon here suggests should follow from such a universal statement, is incorrectly drawn. Mendel is concerned with the laws of *transmission*

* The asymmetries here conceived may of course be combined in an inclusive symmetry. Till the differentiation can be optically recognized in the gametes we shall probably get no further with this part of the problem.

† *Materials for the Study of Variation*, 1894, p. 78.

of existing characters, not with *variation*, which he does not discuss.

Nevertheless Professor Weldon has some acquaintance with the general fact of variability in certain peas, which he mentions (p. 236), but the bearing of this fact on the difficulty he enuntiates escapes him.

Results of crossing in regard to seed characters: normal and exceptional.

The conditions being the same, the question of the characters of the cross-bred zygotes which we will call *AB*'s depends primarily on the specific nature of the varieties which are crossed to produce them. It is unnecessary to point out that if all *AB*'s are to look alike, both the varieties *A* and *B* must be *pure*—not in the common sense of descended, as far as can be traced, through individuals identical with themselves, but pure in the Mendelian sense, that is to say that each must be at that moment producing only homogeneous gametes bearing the same characters *A* and *B* respectively. Purity of pedigree in the breeder's sense is a distinct matter altogether. The length of time—or if preferred—the number of generations through which a character of a variety has remained pure, alters the probability of its *dominance*, i.e. its appearance when a gamete bearing it meets another bearing an antagonistic character, no more, so far as we are yet aware, than the length of time a stable element has been isolated alters the properties of the chemical compound which may be prepared from it.

Now when individuals (bearing contrary characters), pure in the sense indicated, are crossed together, the question arises, What will be the appearance of the first cross individuals? Here again, *generally speaking*, when thoroughly green cotyledons are crossed with thoroughly yellow cotyledons, the first-cross seeds will have yellow cotyledons; when fully round peas are crossed with fully wrinkled the first result will *generally speaking* be *round*, often with slight pitting as Mendel has stated. This has been the usual experience of Correns, Tschermak, Mendel, and myself* and, as we shall see, the amount of clear and substantial evidence to the contrary is still exceedingly small. But as any experienced naturalist would venture to predict, there is no *universal* rule in the matter. As Professor Weldon himself declares, had there been such a universal rule it would surely have been notorious. He might further have reflected that in Mendel's day, when hybridisation was not the *terra incognita* it has since become, the assertion of such universal propositions would have been peculiarly foolish. Mendel does not make it; but Professor Weldon perceiving the inherent improbability of the assertion conceives at once that Mendel *must* have made it, and if Mendel doesn't say so in words then he must have implied it. As a matter of fact Mendel never treats dominance as more than an incident in his results, merely using it as a means to an end, and I see no reason to suppose he troubled to consider to what extent the phenomenon is or is not universal—a matter with which he had no concern.

* The varieties used were *Express*, *Laxton's Alpha*, *Fillbasket*, *McLean's Blue Peter*, *Serpette nain blanc*, *British Queen*, *très nain de Bretagne*, Sabre, *mange-tout* Debarbieux, and a large "grey" sugar-pea, *pois sans parchemin géant à très large cosse*. Not counting the last two, five are round and three are wrinkled. As to cotyledons, six have yellow and four have green. In about 80 crosses I saw no exception to dominance of yellow; but one apparently clear case of dominance of wrinkled and some doubtful ones.

Of course there may be exceptions. As yet we cannot detect the causes which control them, though injury, impurity, accidental crossing, mistakes of various kinds, account for many. Mendel himself says, for instance, that unhealthy or badly grown plants give uncertain results. Nevertheless there seems to be a true residuum of exceptions not to be explained away. I will recite some that I have seen. In my own crosses I have seen green × green give yellow four times. This I incline to attribute to conditions or other disturbance, for the natural pods of these plants gave several yellows. At Messrs Suttons' I saw second-generation seeds got by allowing a cross of *Sutton's Centenary* (gr. wr.) × *Eclipse* (gr. rd.) to go to seed; the resulting seeds were both green and *yellow*, wrinkled and round. But in looking at a sample of *Eclipse* I found a few *yellow* seeds, say two per cent., which may perhaps be the explanation. Green wrinkled × green round *may* give all wrinkled, and again wrinkled × wrinkled may give *round**. Of this I saw a clear case—supposing no mistake to have occurred—at Messrs Suttons'. Lastly we have the fact that in exceptional cases crossing two forms—apparently pure in the strict sense—may give a mixture in the *first* generation. There are doubtless examples also of unlikeness between reciprocals, and of this too I have seen one putative case†.

Such facts thus set out for the first cross-bred generation may without doubt be predicated for subsequent generations.

What then is the significance of the facts?

* Professor Weldon may take this as a famous blow for Mendel, till he realizes what is meant by Mendel's "Hybrid-character."

† In addition to those spoken of later, where the great difference between reciprocals is due to the *maternal* characters of the seeds.

Analysis of exceptions.

Assuming that all these "contradictory" phenomena happened truly as alleged, and were not pathological or due to error—an explanation which seems quite inadequate—there are at least four possible accounts of such diverse results—each valid, without any appeal to ancestry.

1. That dominance may exceptionally fail—or in other words be created on the side which is elsewhere recessive. For this exceptional failure we have to seek exceptional causes. The artificial *creation* of dominance (in a character usually recessive) has not yet to my knowledge been demonstrated experimentally, but experiments are begun by which such evidence may conceivably be obtained.

2. There may be what is known to practical students of evolution as the *false hybridism of Millardet*, or in other words, fertilisation with—from unknown causes—transmission of none or of only some of the characters of one pure parent. The applicability of this hypothesis to the colours and shapes of peas is perhaps remote, but we may notice that it is one possible account of those rare cases where two pure forms give a *mixed* result in the first generation, even assuming the gametes of each pure parent to be truly monomorphic as regards the character they bear. The applicability of this suggestion can of course be tested by study of the subsequent generations, self-fertilised or fertilised by similar forms produced in the same way. In the case of a *genuine* false-hybrid the lost characters will not reappear in the posterity.

3. The result may not be a case of transmission at all as it is at present conceived, but of the creation on crossing

of something *new*. Our *AB*'s may have one or more characters *peculiar to themselves*. We may in fact have made a distinct "mule" or heterozygote form. Where this is the case, there are several subordinate possibilities we need not at present pursue.

4. There may be definite *variation* (distinct from that proper to the "mule") consequent on causes we cannot yet surmise (see pp. 125 and 128).

The above possibilities are I believe at the present time the only ones that need to be considered in connexion with these exceptional cases*. They are all of them capable of experimental test and in certain instances we are beginning to expect the conclusion.

The "mule" or heterozygote.

There can be little doubt that in many cases it is to the third category that the phenomena belong. An indication of the applicability of this reasoning will generally be found in the fact that in such "mule" forms the colour or the shape of the seeds will be recognizably peculiar and proper to the specimens themselves, as distinct from their parents, and we may safely anticipate that when those seeds are grown the plants will show some character which is recognizable as novel. The *proof* that the reasoning may apply can as yet only be got by finding that the forms in

* I have not here considered the case in which male and female elements of a pure variety are not homologous and the variety is a *permanent* monomorphic "mule." Such a phenomenon, when present, will prove itself in reciprocal crossing. I know no such case in peas for certain.

question cannot breed true even after successive selections, but constantly break up into the same series of forms*.

This conception of the "mule" form, or "hybrid-character" as Mendel called it, though undeveloped, is perfectly clear in his work. He says that the dominant character may have two significations, it may be either a parental character or a hybrid-character, and it must be differentiated according as it appears in the one capacity or the other. He does not regard the character displayed by the hybrid, whether dominant or other, *as a thing inherited from or transmitted by the pure parent at all, but as the peculiar function or property of the hybrid.* When this conception has been fully understood and appreciated in all its bearings it will be found to be hardly less fruitful than that of the purity of the germ-cells.

The two parents are two—let us say—substances† represented by corresponding gametes. These gametes unite to form a new "substance"—the cross-bred zygote. This has its own properties and structure, just as a chemical compound has, and the properties of this new "substance" are *not more strictly* traceable to, or "inherited" from, those of the two parents than are those of a new chemical compound "inherited" from those of the component elements. If the case be one in which the gametes are pure, the new "substance" is not represented by them, but the compound is again dissociated into its components, each of which is separately represented by gametes.

* It will be understood that a "mule" form is quite distinct from what is generally described as a "blend." One certain criterion of the "mule" form is the fact that it cannot be fixed, see p. 25. There is little doubt that Laxton had such a "mule" form when he speaks of "the remarkably fine but unfixable pea, Evolution." *J. R. Hort. Soc.* XII. 1890, p. 37 (*v. infra*).

† Using the word metaphorically.

The character of the cross-bred zygote may be anything. It may be something we have seen before in one or other of the parents, it may be intermediate between the two, or it may be something new. All these possibilities were known to Mendel and he is perfectly aware that his principle is equally applicable to all. The first case is his "dominance." That he is ready for the second is sufficiently shown by his brief reference to time of flowering considered as a character (p. 65). The hybrids, he says, flower at a time *almost exactly intermediate* between the flowering times of the parents, and he remarks that the development of the hybrids in this case probably happens in the same way as it does in the case of the other characters*.

That he was thoroughly prepared for the third possibility appears constantly through the paper, notably in the argument based on the *Phaseolus* hybrids, and in the statement that the hybrid between talls and dwarfs is generally taller than the tall parent, having increased height as its "hybrid-character."

All this Professor Weldon has missed. In place of it he offers us the *sententia* that no one can expect to understand these phenomena if he neglect ancestry. This is the idle gloss of the scribe, which, if we erase it not thoroughly, may pass into the text.

Enough has been said to show how greatly Mendel's conception of heredity was in advance of those which pass current at the present day; I have here attempted

* "*Ueber die Blüthezeit der Hybriden sind die Versuche noch nicht abgeschlossen. So viel kann indessen schon angegeben werden, dass dieselbe fast genau in der Mitte zwischen jener der Samen- und Pollenpflanze steht, und die Entwicklung der Hybriden bezüglich dieses Merkmales wahrscheinlich in der nämlichen Weise erfolgt, wie es für die übrigen Merkmale der Fall ist.*" Mendel, p. 23.

the barest outline of the nature of the "hybrid-character," and I have not sought to indicate the conclusions that we reach when the reasoning so clear in the case of the hybrid is applied to the pure forms and their own characters.

In these considerations we reach the very base on which all conceptions of heredity and variation must henceforth rest, and that it is now possible for us to attempt any such analysis is one of the most far-reaching consequences of Mendel's principle. Till two years ago no one had made more than random soundings of this abyss.

I have briefly discussed these possibilities to assist the reader in getting an insight into Mendel's conceptions. But in dealing with Professor Weldon we need not make this excursion; for his objection arising from the absence of uniform regularity in dominance is not in point.

The soundness of Mendel's work and conclusions would be just as complete if dominance be found to fail often instead of rarely. For it is perfectly certain that varieties *can* be chosen in such a way that the dominance of one character over its antagonist is so regular a phenomenon that it *can* be used in the way Mendel indicates. He chose varieties, in fact, in which a known character *was* regularly dominant and it is because he did so that he made his discovery*. When Professor Weldon speaks of the existence of fluctuation and diversity in regard to dominance as proof of a "grave discrepancy" between Mendel's facts and those of other observers†, he merely indicates the point at which his own misconceptions began.

* As has been already shown the discovery could have been made equally well and possibly with greater rapidity in a case in which the hybrid had a character distinct from either parent. The cases that would *not* have given a clear result are those where there is irregular dominance of one or other parent.

† Weldon, p. 240.

From Mendel's style it may be inferred that if he had meant to state universal dominance in peas he would have done so in unequivocal language. Let me point out further that of the 34 varieties he collected for study, he discarded 12 as not amenable to his purposes*. He tells us he would have nothing to do with characters which were not sharp, but of a "more or less" description. As the 34 varieties are said to have all come true from seed, we may fairly suppose that the reason he discarded twelve was that they were unsuitable for his calculations, having either ill-defined and intermediate characters, or possibly defective and irregular dominance.

IV. Professor Weldon's collection of "Other Evidence concerning Dominance in Peas."

A. In regard to cotyledon colour: Preliminary.

I have been at some pains to show how the contradictory results, no doubt sometimes occurring, on which Professor Weldon lays such stress, may be comprehended without any injury to Mendel's main conclusions. This excursion was made to save trouble with future discoverers of exceptions, though the existence of such facts need scarcely disturb many minds. As regards the dominance of yellow cotyledon-colour over green the whole number of genuine unconformable cases is likely to prove very small indeed, though in regard to the dominance of round shape over wrinkled we may be prepared for more discrepancies. Indeed my own crosses alone are sufficient to show that in using some varieties irregularities are to be expected.

* See p. 43.

Considering also that the shapes of peas depend unquestionably on more than one pair of allelomorphs I fully expect regular blending in some cases.

As however it may be more satisfactory to the reader and to Professor Weldon if I follow him through his "contradictory" evidence I will endeavour to do so. Those who have even a slight practical acquaintance with the phenomena of heredity will sympathize with me in the difficulty I feel in treating this section of his arguments with that gravity he conceives the occasion to demand.

In following the path of the critic it will be necessary for me to trouble the reader with a number of details of a humble order, but the journey will not prove devoid of entertainment.

Now exceptions are always interesting and suggestive things, and sometimes hold a key to great mysteries. Still when a few exceptions are found disobeying rules elsewhere conformed to by large classes of phenomena it is not an unsafe course to consider, with such care as the case permits, whether the exceptions may not be due to exceptional causes, or failing such causes whether there may be any possibility of error. But to Professor Weldon, an exception is an exception—and as such may prove a very serviceable missile; so he gathers them as they were "smooth stones from the brook."

Before examining the quality of this rather miscellaneous ammunition I would wish to draw the non-botanical reader's attention to one or two facts of a general nature.

For our present purpose the seed of a pea may be considered as consisting of two parts, the *embryo with its cotyledons*, enclosed in a *seed-coat*. It has been known for about a century that this coat or skin is a *maternal* structure, being part of the mother plant just as much as the pods are, and consequently not belonging to the next generation at all. If then any changes take place in it consequent on fertilisation, they are to be regarded not as in any sense a transmission of character by heredity, but rather as of the nature of an "infection." If on the other hand it is desired to study the influence of hereditary transmission on seed-coat characters, then the crossed seeds must be sown and the seed-coats of their seeds studied. Such infective changes in maternal tissues have been known from early times, a notable collection of them having been made especially by Darwin; and for these cases Focke suggested the convenient word *Xenia.* With this familiar fact I would not for a moment suppose Professor Weldon unacquainted, though it was with some surprise that I found in his paper no reference to the phenomenon.

For as it happens, xenia is not at all a rare occurrence with *certain varieties* of peas; though in them, as I believe is generally the case with this phenomenon, it is highly irregular in its manifestations, being doubtless dependent on slight differences of conditions during ripening.

The coats of peas differ greatly in different varieties, being sometimes thick and white or yellow, sometimes thick and highly pigmented with green or other colours, in both of which cases it may be impossible to judge the cotyledon-colour without peeling off the opaque coat; or the coats may be very thin, colourless and transparent, so that the cotyledon-colour is seen at once. It was such a transparent form that Mendel says he used for his experiments with cotyledon-colour. In order to see xenia a pea with a *pigmented* seed-coat should be taken as seed-parent, and crossed with a variety having a different cotyledon-colour. There is then a fair chance of seeing this phenomenon, but much still depends on the variety. For example, *Fillbasket* has green cotyledons and seed-coat green except near the hilar surface. Crossed with *Serpette nain blanc* (yellow cotyledons and yellow coat) this variety gave three pods with 17 seeds in which the seed-coats were almost full yellow (xenia). Three other pods (25 seeds), similarly produced, showed slight xenia, and one pod with eight seeds showed little or none.

On the other hand *Fillbasket* fertilised with *nain de Bretagne* (yellow cotyledons, seed-coats yellow to yellowish green) gave six pods with 39 seeds showing slight xenia, distinct in a few seeds but absent in most.

Examples of xenia produced by the contrary proceeding, namely fertilising a yellow pea with a green, may indubitably occur and I have seen doubtful cases ; but as by the nature of the case these are *negative* phenomena, i.e. the seed-coat remaining greenish and *not* going through its normal maturation changes, they must always be equivocal, and would require special confirmation before other causes were excluded.

Lastly, the special change (xenia) Mendel saw in "grey" peas, appearance or increase of purple pigment in the thick coats, following crossing, is common but also irregular.

If a *transparent* coated form be taken as seed-parent there is no appreciable xenia, so far as I know, and such a phenomenon would certainly be paradoxical*.

In this connection it is interesting to observe that Giltay, whom Professor Weldon quotes as having obtained purely Mendelian results, got no xenia though searching for it. If the reader goes carefully through Giltay's numerous cases, he will find, *almost* without doubt, that none of them were such as produce it. *Reading Giant*, as

* In some transparent coats there is pigment, but so little as a rule that xenia would be scarcely noticeable.

Giltay states, has a *transparent* skin, and the only xenia likely to occur in the other cases would be of the peculiar and uncertain kind seen in using "grey" peas. Professor Weldon notes that Giltay, who evidently worked with extreme care, *peeled* his seeds before describing them, a course which Professor Weldon, not recognizing the distinction between the varieties with opaque and transparent coats, himself wisely recommends. The coincidence of the peeled seeds giving simple Mendelian results is one which might have alarmed a critic less intrepid than Professor Weldon.

Bearing in mind, then, that the coats of peas may be transparent or opaque; and in the latter case may be variously pigmented, green, grey, reddish, purplish, etc.; that in any of the latter cases there may or may not be xenia; the reader will perceive that to use the statements of an author, whether scientific or lay, to the effect that on crossing varieties he obtained peas of such and such colours *without specifying at all whether the coats were transparent or whether the colours he saw were coat- or cotyledon-colours* is a proceeding fraught with peculiar and special risks.

(1) *Gärtner's cases.* Professor Weldon gives, as exceptions, a series of Gärtner's observations. Using several varieties, amongst them *Pisum sativum macrospermum*, a "grey" pea, with coloured flowers and seed-coats*, he obtained results partly Mendelian and partly, as now alleged, contradictory. The latter consist of seeds "dirty yellow" and "yellowish green," whereas it is suggested they should have been simply yellow.

Now students of this department of natural history will know that these same observations of Gärtner's, whether rightly or wrongly, have been doing duty for more than half a century as stock illustrations of xenia. In this

* Usually correlated characters, as Mendel knew.

capacity they have served two generations of naturalists. The ground nowadays may be unfamiliar, but others have travelled it before and recorded their impressions. Darwin, for example, has the following passage*:

"These statements led Gärtner, who was highly sceptical on the subject, carefully to try a long series of experiments; he selected the most constant varieties, and the results conclusively showed *that the colour of the skin of the pea* is modified when pollen of a differently coloured variety is used." (The italics are mine.)

In the true spirit of inquiry Professor Weldon doubtless reflected,

"'Tis not *Antiquity* nor *Author*,
That makes *Truth Truth*, altho' *Time's Daughter*";

but perhaps a word of caution to the reader that another interpretation exists would have been in place. It cannot be without amazement therefore that we find him appropriating these examples as referring to cotyledon-colour, with never a hint that the point is doubtful.

Giltay, without going into details, points out the ambiguity†. As Professor Weldon refers to the writings both of Darwin and Giltay, it is still more remarkable that he should regard the phenomenon as clearly one of cotyledon-colour and not coat-colour as Darwin and many other writers have supposed.

* *Animals and Plants*, 2nd ed. 1885, p. 428.

† "*Eine andere Frage ist jedoch, ob der Einfluss des Pollens auf den Keim schon äusserlich an diesen letzteren sichtbar sein kann. Darwin führt mehrere hierher gehörige Fälle an, und wahrscheinlich sind auch die Resultate der von Gärtner über diesen Gegenstand ausgeführten Experimente hier zu erwähnen, wenn es auch nicht ganz deutlich ist, ob der von Gärtner erwähnte directe Einfluss des Pollens sich nur innerhalb der Grenzen des Keimes merklich macht oder nicht.*" p. 490.

Without going further it would be highly improbable that Gärtner is speaking solely or even chiefly of the cotyledons, from the circumstance that these observations are given as evidence of "*the influence of foreign pollen on the female organs*"; and that Gärtner was perfectly aware of the fact that the coat of the seed was a maternal structure is evident from his statement to that effect on p. 80.

To go into the whole question in detail would require considerable space; but indeed it is unnecessary to labour the point. The reader who examines Gärtner's account with care, especially the peculiar phenomena obtained in the case of the "grey" pea (*macrospermum*), with specimens before him, will have no difficulty in recognizing that Gärtner is simply describing the seeds *as they looked in their coats*, and is not attempting to distinguish cotyledon-characters and coat-characters. If he had peeled them, which in the case of "grey" peas would be *absolutely necessary* to see cotyledon-colour, he must surely have said so.

Had he done so, he would have found the cotyledons full yellow in every ripe seed; for I venture to assert that anyone who tries, as we have, crosses between a yellow-cotyledoned "grey" pea, such as Gärtner's was, with any pure green variety will see that there is no question whatever as to absolute dominance of the yellow cotyledon-character here, more striking than in any other case. If exceptions are to be looked for, they will not be found *there*; and, except in so far as they show simple dominance of yellow, Gärtner's observations cannot be cited in this connection at all.

(2) *Seton's case.* Another exception given by Professor Weldon is much more interesting and instructive.

It is the curious case of Seton*. Told in the words of the critic it is as follows :—

"Mr Alexander Seton crossed the flowers of *Dwarf Imperial*, 'a well-known green variety of the Pea,' with the pollen of 'a white free-growing variety.' Four hybrid seeds were obtained, 'which did not differ in appearance from the others of the female parent.' These seeds therefore did *not* obey the law of dominance, or if the statement be preferred, greenness became dominant in this case. The seeds were sown, and produced plants bearing 'green' and 'white' seeds side by side in the same pod. An excellent coloured figure of one of these pods is given (*loc. cit.* Plate 9, Fig. 1), and is the only figure I have found which illustrates segregation of colours in hybrid Peas of the second generation."

Now if Professor Weldon had applied to this case the same independence of judgment he evinced in dismissing Darwin's interpretation of Gärtner's observations, he might have reached a valuable result. Knowing how difficult it is to give all the points in a brief citation, I turned up the original passage, where I find it stated that the mixed seeds of the second generation "were all completely either of one colour or the other, none of them having an intermediate tint, as Mr Seton had expected." The utility of this observation of the absence of intermediates, is that it goes some way to dispose of the suggestion of xenia as a cause contributing to the result.

Moreover, feeling perfectly clear, from the fact of the absence of intermediates, that the case must be one of simple dominance in spite of first appearances, I suggest the following account with every confidence that it is the true one. There have been several "*Imperials*,"

* Appendix to paper of Goss, *Trans. Hort. Soc.* v. 1822, pub. 1824 (*not* 1848, as given by Professor Weldon), p. 236.

though *Dwarf Imperial*, in a form which I can feel sure is Seton's form, I have not succeeded in seeing; but from Vilmorin's description that the peas when ripe are "*franchement verts*" I feel no doubt it was a green pea *with a green skin*. If it had had a transparent skin this description would be inapplicable. Having then a green skin, which may be assumed with every probability of truth, the seeds, even though the cotyledons were yellow, might, especially if examined fresh, be indistinguishable from those of the maternal type. Next from the fact of the mixture in the second generation we learn that the *semi-transparent seed-coat of the paternal form was dominant* as a plant-character, and indeed the coloured plate makes this fairly evident. It will be understood that this explanation is as yet suggestive, but from the facts of the second generation, any supposition that there was real irregularity in dominance in this case is out of the question*.

(3) *Tschermak's exceptions*. These are a much more acceptable lot than those we have been considering. Tschermak was thoroughly alive to the seed-coat question and consequently any exception stated as an unqualified fact on his authority must be accepted. The nature of these cases we shall see. Among the many varieties he used, some being *not* monomorphic, it would have been surprising if he had not found true irregularities in dominance.

(3 *a*) *Buchsbaum case*. This variety, growing in the open, gave once a pod in which *every seed but one was green*. In stating this case Professor Weldon refers to *Buchsbaum*

* Since the above passage was written I find the "*Imperials*" described in "Report of Chiswick Trials," *Proc. R. Hort. Soc.* 1860, I. p. 340, as "skin thick"; and on p. 360 "skin thick, blue"; which finally disposes of this "exception."

as "a yellow-seeded variety." Tschermak*, however, describes it as having "*gelbes, öfters gelblich-grünes Speichergewebe*" (cotyledons); and again says the cotyledon-colour is "*allerdings gerade bei Buchsbaum zur Spontanvariation nach gelb-grün neigend!*" The (!) is Tschermak's. Therefore Professor Weldon can hardly claim *Buchsbaum* as "yellow-seeded" without qualification.

Buchsbaum in fact is in all probability a blend-form and certainly not a true, stable yellow. One of the green seeds mentioned above grew and gave 15 *yellows* and three *greens,* and the result showed pretty clearly, as Tschermak says, that there had been an accidental cross with a tall green.

On another occasion *Telephone* ♀ (another impure green) × *Buchsbaum* gave four *yellow smooth and* two *green wrinkled,* but one [? both: the grammar is obscure] of the greens did not germinate†.

(3 *b*) *Telephone cases. Telephone,* crossed with at least one yellow variety (*Auvergne*) gave all or some green or greenish. These I have no doubt are good cases of "defective dominance" of yellow. But it must be noted that *Telephone is an impure green.* Nominally a green, it is as Professor Weldon has satisfied himself, very irregular in colour, having many intermediates shading to pure yellow and many piebalds. It is the variety from which alone Professor Weldon made his colour-scale. *I desire therefore to call special attention to the fact that Telephone, though*

* (36), p. 502 and (37), p. 663.

† Professor Weldon should have alluded to this. *Dead* seeds have no bearing on these questions, seeing that their characters may be pathological. The same seeds are later described as "*wie Telephone selbst,*" so, apart from the possibility of death, they may also have been self-fertilised.

not a pure green, Tschermak's sample being as he says "gelblichweiss grün," a yellowish-white-green in cotyledon-colour, is the variety which has so far contributed the clearest evidence of the green colour dominating in its crosses with a yellow; and that *Buchsbaum* is probably a similar case. To this point we shall return. It may not be superfluous to mention also that one cross between *Fillbasket* (a thorough *green*) and *Telephone* gave three *yellowish* green seeds (Tschermak, (36), p. 501).

(3 *c*) *Couturier cases.* This fully yellow variety in crosses with two fully green sorts gave seeds either yellow or greenish yellow. In one case *Fillbasket* ♀ fertilised by *Couturier* gave mixed seeds, green and yellow. For any evidence to the contrary, the green in this case may have been self-fertilised. Nevertheless, taking the evidence together, I think it is most likely that *Couturier* is a genuine case of imperfect dominance of yellow. If so, it is the only true "exception" in crosses between stable forms.

We have now narrowed down Professor Weldon's exceptions to dominance of cotyledon-colour to two varieties, one yellow (*Couturier*), and one yellow "tending to green" (*Buchsbaum*), which show imperfect dominance of *yellow*; and one variety, *Telephone*, an impure and irregular green, which shows occasional but uncertain dominance of *green*.

What may be the meaning of the phenomenon shown by the unstable or mosaic varieties we cannot tell; but I venture to suggest that when we more fully appreciate the nature and genesis of the gametes, it will be found that the peculiarities of heredity seen in these cases have more in common with those of "false hybridism" (see p. 34) than with any true failure of dominance.

Before, however, feeling quite satisfied in regard even

to this residuum of exceptions, one would wish to learn the subsequent fate of these aberrant seeds and how their offspring differed from that of their sisters. One only of them can I yet trace, viz. the green seed from *Telephone* ♀ × *Buchsbaum* ♂, which proved a veritable "green dominant." As for the remainder, Tschermak promises in his first paper to watch them. But in his second paper the only passage I can find relating to them declares that perhaps some of the questionable cases he mentioned in his first paper "*are attributable to similar isolated anomalies in dominance; some proved themselves by subsequent cultivation to be cases of accidental self-fertilisation; others failed to germinate**." I may warn those interested in these questions, that in estimating changes due to ripening, *dead* seeds are not available.

B. Seed-coats and shapes.

1. *Seed-coats.* Professor Weldon lays some stress on the results obtained by Correns† in crossing a pea having green cotyledons and a thin almost colourless coat (*grüne späte Erfurter Folger-erbse*) with two purple-flowered varieties. The latter are what are known in England as "grey" peas, though the term grey is not generally appropriate.

In these varieties the cotyledon-colour is yellow and

* "*Vielleicht sind einige der l.c.* 507 *bis* 508 *erwähnten fraglichen Fälle auf ähnliche vereinzelte Anomalien der Merkmalswerthigkeit zu beziehen; einige erwiesen sich allerdings beim Anbau als Producte ungewollter Selbstbefruchtung, andere keimten nicht.*"

† Regarding this case I have to thank Professor Correns for a good deal of information which he kindly sent me in response to my inquiry. I am thus able to supplement the published account in some particulars.

the coats are usually highly coloured or orange-brown. In reciprocal crosses Correns found no change from the maternal seed-coat-colour or seed-shape. On sowing these peas he obtained plants bearing peas which, using the terminology of Mendel and others, he speaks of as the "first generation."

These peas varied in the colour of their seed-coats from an almost colourless form slightly tinged with green like the one parent to the orange-brown of the other parent. The seeds varied in this respect not only from plant to plant, but from pod to pod, and from seed to seed, as Professor Correns has informed me.

The peas with more highly-coloured coats were sown and gave rise to plants with seeds showing the whole range of seed-coat-colours again.

Professor Weldon states that in this case neither the law of dominance nor the law of segregation was observed; and the same is the opinion of Correns, who, as I understand, inclines to regard the colour-distribution as indicating a "mosaic" formation. This is perhaps conceivable; and in that case the statement that there was no dominance would be true, and it would also be true that the unit of segregation, if any, was smaller than the individual plant and may in fact be the individual seed.

A final decision of this question is as yet impossible. Nevertheless from Professor Correns I have learnt one point of importance, namely, that the coats of all these seeds were *thick*, like that of the coloured and as usual dominant form. There is no "mosaic" of coats like one parent and coats like the other, though there may be a mosaic of colours. In regard to the distribution of *colour* however the possibility does not seem to me excluded that we are here dealing with changes influenced by conditions.

I have grown a "grey" pea and noticed that the seed-coats ripened in my garden differ considerably and not quite uniformly from those received from and probably ripened in France, mine being mostly pale and greyish, instead of reddish-brown. We have elsewhere seen (p. 120) that pigments of the seed-coat-colour may be very sensitive to conditions, and slight differences of moisture, for example, may in some measure account for the differences in colour. Among my crosses I have a pod of such "grey" peas fertilised by *Laxton's Alpha* (green cotyledons, coat transparent). It contained five seeds, of which four were *red-brown on one side* and grey with purple specks on the other. The fifth was of the grey colour on both sides. I regard this difference not as indicating segregation of character but merely as comparable with the difference between the two sides of a ripe apple, and I have little doubt that Correns' case may be of the same nature*. Phenomena somewhat similar to these will be met with in Laxton's case of the "maple" seeded peas (see p. 161).

2. *Seed-shapes.* Here Professor Weldon has three sets of alleged exceptions to the rule of dominance of round shape over wrinkled. The first are Rimpau's cases, the second are Tschermak's cases, the third group are cases of "grey" peas, which we will treat in a separate section (see pp. 153 and 158).

(*a*) *Rimpau's cases.* Professor Weldon quotes Rimpau as having crossed wrinkled and round peas† and found

* Mr Hurst, of Burbage, tells me that in varieties having coats green or white, e.g. *American Wonder*, the white coats are mostly from early, the green from later pods, the tints depending on conditions and exposure.

† In the first case *Knight's Marrow* with *Victoria*, both ways; in the second *Victoria* with *Telephone*, both ways.

the second hybrid generation dimorphic as usual. The wrinkled peas were selected and sown and gave wrinkled peas *and round* peas, becoming "true" to the wrinkled character in one case only in the fifth year, while in the second case—that of a *Telephone* cross—there was a mixture of round and wrinkled similarly resulting from *wrinkled* seed for two years, but the experiment was not continued.

These at first sight look like genuine exceptions. In reality, however, they are capable of a simple explanation. It must be remembered that Rimpau was working in ignorance of Mendel's results, was not testing any rule, and was not on the look out for irregularities. Now all who have crossed wrinkled and round peas on even a moderate scale will have met with the fact that there is frequently *some* wrinkling in the cross-bred seeds. Though round when compared with the true wrinkled, these are often somewhat more wrinkled than the round type, and in irregular degrees. For my own part I fully anticipate that we may find rare cases of complete blending in this respect though I do not as yet know one.

Rimpau gives a photograph of eight peas (Fig. 146) which he says represent the wrinkled form derived from this cross. It is evident that these are not from *one pod* but a miscellaneous selection. On close inspection it will be seen that while the remainder are shown with their *cotyledon*-surfaces upwards, the two peas at the lower end of the row are represented with their *hilar*-surfaces upwards. Remembering this it will be recognized that these two lower peas are in fact *not* fully wrinkled peas but almost certainly *round* "hybrids," and the depression is merely that which is often seen in round peas (such as *Fillbasket*), squared by mutual pressure. Such peas, when sown, might of course give some round.

As Tschermak writes ((37), p. 658), experience has shown him that cross-bred seeds with character transitional between "round" and "wrinkled" behave as hybrids, and have both wrinkled and round offspring, and he now reckons them accordingly with the round dominants.

Note further the fact that Rimpau found the wrinkled form came true in the *fifth* year, while the round gave at first more, later fewer, wrinkleds, not coming true till the *ninth* year. This makes it quite clear that there *was* dominance of the round form, but that the heterozygotes were not so sharply distinguishable from the two pure forms as to be separated at once by a person not on the look-out for the distinctions. Nevertheless there *was* sufficient difference to lead to a practical distinction of the cross-breds both from the pure dominants and from the pure recessives.

The *Telephone* case may have been of the same nature; though, as we have seen above, this pea is peculiar in its colour-heredity and may quite well have followed a different rule in shape also. As stated before, the wrinkled offspring were not cultivated after the third year, but the *round* seeds are said to have still given some wrinkleds in the eighth year after the cross, as would be expected in a simple Mendelian case.

(*b*) *Tschermak's cases.* The cases Professor Weldon quotes from Tschermak all relate to crosses with *Telephone* again, and this fact taken with the certainty that the colour-heredity of *Telephone* is abnormal makes it fairly clear that there is here something of a really exceptional character. What the real nature of the exception is, and how far it is to be taken as contradicting the "law of dominance," is quite another matter.

3. *Other phenomena, especially regarding seed-shapes, in the case of "grey" peas. Modern evidence.* Professor Weldon quotes from Tschermak the interesting facts about the "grey" pea, *Graue Riesen*, but does not attempt to elucidate them. He is not on very safe ground in adducing these phenomena as conflicting with the "law of dominance." Let us see whither we are led if we consider these cases. On p. 124 I mentioned that the classes round and wrinkled do not properly hold if we try to extend them to large-seeded sorts, and that these cases require separate consideration. In many of such peas, which usually belong either to the classes of sugar-peas (*mange-touts*) or "grey" peas (with coloured flowers), the seeds would be rather described as irregularly indented, lumpy or stony*, than by any use of the terms round or wrinkled. One sugar-pea (*Debarbieux*) which I have used has large flattish, smooth, yellow seeds with white skins, and this also in its crossings follows the rules about to be described for the large-seeded "grey" peas.

In the large "grey" peas the most conspicuous feature is the seed-coat, which is grey, brownish, or of a bright reddish colour. Such seed-coats are often speckled with purple, and on boiling these seed-coats turn dark brown. They are in fact the very peas used by Mendel in making up his third pair of characters. Regarding them Professor

* Gärtner's *macrospermum* was evidently one of these, though from the further account (p. 498) it was probably more wrinkled. There are of course *mange-touts* which have perfectly round seeds. Mendel himself showed that the *mange-tout* character, the soft constricted pod, was transferable. There are also *mange-touts* with fully wrinkled seeds and "grey" peas with small seeds (see Vilmorin-Andrieux, *Plantes Potagères*, 1883).

Weldon, stating they may be considered separately, writes as follows:—

"Tschermak has crossed *Graue Riesen* with five races of *P. sativum*, and he finds that the form of the first hybrid seeds *follows the female parent*, so that if races of *P. sativum* with round smooth seeds be crossed with *Graue Riesen* (which has flattened, feebly wrinkled seeds) the hybrids will be round and smooth or flattened and wrinkled, as the *P. sativum* or the *Graue Riesen* is used as female parent*. There is here a more complex phenomenon than at first sight appears; because if the flowers of the first hybrid generation are self-fertilised, the resulting seeds of the second generation invariably resemble those of the *Graue Riesen* in shape, although in colour they follow Mendel's law of segregation!"

From this account who would not infer that we have here some mystery which does not accord with the Mendelian principles? As a matter of fact the case is dominance in a perfectly obvious if distinct form.

Graue Riesen, a large grey sugar-pea, the *pois sans parchemin géant* of the French seedsmen, has full-yellow cotyledons and a highly coloured seed-coat of varying tints. In shape the seed is somewhat flattened with irregular slight indentations, lightly wrinkled if the term be preferred. Tschermak speaks of it in his first paper as "*Same flach, zusammengedrückt*"—a flat, compressed seed; in his second paper as "*flache, oft schwach gerunzelte Cotyledonen-form,*" or cotyledon-shape, flat, often feebly wrinkled, as Professor Weldon translates.

First-crosses made from this variety, each with a different form of *P. sativum*, are stated on the authority of Tschermak's five cases, to follow exclusively the maternal seed-shape. From "*schwach gerunzelte,*" "feebly wrinkled," Professor Weldon easily passes to "wrinkled," and tells us

* Correns found a similar result.

that according as a round *sativum* or the *Graue Riesen* is used as mother, the first-cross seeds "will be round and smooth or flattened and wrinkled."

As a matter of fact, however, the seeds of *Graue Riesen* though *slightly* wrinkled do not belong to the "wrinkled" class; but if the classification "wrinkled" and "round" is to be extended to such peas at all, they belong to the *round*. Mendel is careful to state that his *round* class are "either spherical or roundish, the depressions on the surface, when there are any, always slight"; while the "wrinkled" class are "irregularly angular, deeply wrinkled*."

On this description alone it would be very likely that *Graue Riesen* should fall into the *round* class, and as such it behaves in its crosses, *being dominant over wrinkled* (see Nos. 3 and 6, below). I can see that in this case Professor Weldon has been partly misled by expressions of Tschermak's, but the facts of the second generation should have aroused suspicion. Neither author notices that as all five varieties crossed by Tschermak with *Graue Riesen* were *round*, the possibilities are not exhausted. Had Tschermak tried a really wrinkled *sativum* with *Graue Riesen* he would have seen this obvious explanation.

As some of my own few observations of first-crosses bear on this point I may quote them, imperfect though they are.

I grew the purple-flowered sugar-pea "*Pois sans parchemin géant à très large cosse*," a soft-podded "*mangetout*" pea, flowers and seed-coats coloured, from Vilmorin's, probably identical with *Graue Riesen*.

1. One flower of this variety fertilised with *Pois très nain de Bretagne* (very small seed; yellow cotyledons; very

* "*Entweder kugelrund oder rundlich, die Einsenkungen, wenn welche an der Oberfläche vorkommen, immer nur seicht, oder sie sind unregelmässig kantig, tief runzlig* (*P. quadratum*)."

round) gave seven seeds indistinguishable (in their coats) from those of the mother, save for a doubtful increase in purple pigmentation of coats.

2. Fertilised by *Laxton's Alpha* (green; wrinkled; coats transparent), two flowers gave 11 seeds exactly as above, the purple being in this case clearly increased.

In the following the purple sugar-pea was *father*.

3. *Laxton's Alpha* (green; wrinkled; coats transparent) fertilised by the purple sugar-pea gave one pod of four seeds with yellow cotyledons and *round* form.

4. *Fillbasket* (green; smooth but squared; coats green) fertilised by the *purple* sugar-pea gave one pod with six seeds, yellow cotyledons*; *Fillbasket* size and shape; but the normally green coat yellowed near *the hilum* by xenia.

5. *Express* ("blue"-green cotyledons and transparent skins; round) fertilised with *purple sugar*-pea gave one pod with four seeds, yellow cotyledons, shape round, much as in *Fillbasket*.

6. *British Queen* (yellow cotyledons, wrinkled, white coats) ♀ × purple sugar-pea gave two pods with seven seeds, cotyledons yellow, coats *tinged greenish* (xenia?), all *round*.

So much for the "*Purple*" sugar-pea.

I got similar results with *Mange-tout Debarbieux*. This is a soft-podded *Mange-tout* or sugar-pea, with white flowers, large, flattish, smooth seeds, scarcely dimpled; yellow cotyledons.

* The colour is the peculiarly deep yellow of the "grey" *mange-tout*.

7. *Debarbieux* fertilised by *Serpette nain blanc* (yellow cotyledons; wrinkled; white skin; dwarf) gave one pod with six seeds, size and shape of *Debarbieux*, with slight dimpling.

8. *Debarbieux* by *nain de Bretagne* (very small; yellow cotyledons; very round) gave three pods, 12 seeds, all yellow cotyledons, of which two pods had eight seeds identical in shape with *Debarbieux*, while the third had four seeds like *Debarbieux* but more dimpled. The reciprocal cross gave two seeds exactly like *nain de Bretagne*.

But it may be objected that the shape of this large grey pea is very peculiar* ; and that it maintains its type remarkably when fertilised by many distinct varieties though its pollen effects little or no change in them ; for, so long as round varieties of *sativum* are used as mothers, this is true as we have seen. But when once it is understood that in *Graue Riesen* there is no question of wrinkling, seeing that the variety behaves as a *round* variety, the shape and especially the size of the seed must be treated as a maternal property.

Why the distinction between the shape of *Graue Riesen* and that of ordinary round peas should be a matter of maternal physiology we do not know. The question is one for the botanical chemist. But there is evidently very considerable regularity, the seeds borne by the *cross-breds* exhibiting the form of the "grey" pea, which is then a dominant character as much as the seed-coat characters

* It is certainly subject to considerable changes according to conditions. Those ripened in my garden are without exception much larger and flatter than Vilmorin's seeds (now two years old) from which they grew. The colour of the coats is also much duller. These changes are just what is to be expected from the English climate—taken with the fact that my sample of this variety was late sown.

are. And that is what Tschermak's *Graue Riesen* crosses actually did, thereby exhibiting dominance in a very clear form. To interject these cases as a mystery without pointing out how easily they can be reconciled with the "law of dominance" may throw an unskilled reader into gratuitous doubt.

Finally, since *the wrinkled peas, Laxton's Alpha* and *British Queen, pollinated by a large flat mange-tout, witness Nos.* 3 *and* 6 *above*, became round in both cases where this experiment was made, we here merely see the usual dominance of the non-wrinkled character; though of course if a *round*-seeded mother be used there can be no departure from the maternal shape, as far as roundness is concerned.

Correns' observations on the shapes of a "grey" pea crossed with a round shelling pea, also quoted by Professor Weldon as showing no dominance of roundness, are of course of the same nature as those just discussed.

C. Evidence of Knight and Laxton.

In the last two sections we have seen that in using peas of the "grey" class, i.e. with brown, red, or purplish coats, special phenomena are to be looked for, and also that in the case of large "indented" peas, the phenomena of size and shape may show some divergence from that simple form of the phenomenon of dominance seen when ordinary round and wrinkled are crossed. Here the fuller discussion of these phenomena must have been left to await further experiment, were it not that we have other evidence bearing on the same questions.

The first is that of Knight's well-known experiments, long familiar but until now hopelessly mysterious. I have not space to quote the various interpretations which Knight and others have put upon them, but as the Mendelian principle at once gives a complete account of the whole, this is scarcely necessary, though the matter is full of historical interest.

Crossing a white pea with a very large grey purple-flowered form Knight (21) found that the peas so produced "were not in any sensible degree different from those afforded by other plants of the same [white] variety; owing, I imagine, to the external covering of the seed (as I have found in other plants) being furnished entirely by the female*." All grew very tall†, and had colours of male parent‡. The seeds they produced were dark grey§.

"I had frequent occasion to observe, in this plant [the hybrid], a stronger tendency to produce purple blossoms, and coloured seeds, than white ones; for when I introduced the farina of a purple blossom into a white one, the whole of the seeds in the succeeding year became coloured [viz. $DR \times D$ giving DD and DR]; but, when I endeavoured to discharge this colour, by reversing the process, a part only of them afforded plants with white blossoms; this part sometimes occupying one end of the pod, and being at times irregularly intermixed with those which, when sown, retained their colour" [viz. $DR \times R$ giving DR and RR] (draws conclusions, now obviously erroneous‖).

In this account we have nothing not readily intelligible in the light of Mendel's hypothesis.

The next evidence is supplied by an exceptionally complete record of a most valuable experiment made by

* Thus avoiding the error of Seton, see p. 144. There is no xenia perhaps because the seed-coat of mother was a transparent coat.

† As heterozygotes often do.

‡ Dominance of the purple form.

§ Dominance of the grey coat as a maternal character.

‖ Sherwood's view (*J. R. Hort. Soc.* XXII. p. 252) that this was the origin of the "Wrinkled" pea, seems very dubious.

Laxton*. The whole story is replete with interest, and as it not only carries us on somewhat beyond the point reached by Mendel, but furnishes an excellent illustration of how his principles may be applied, I give the whole account in Laxton's words, only altering the paragraphing for clearness, and adding a commentary. The paper appears in *Jour. Hort. Soc.* N.S. III. 1872, p. 10, and very slightly abbreviated in *Jour. of Hort.* XVIII. 1870, p. 86. Some points in the same article do not specially relate to this section, but for simplicity I treat the whole together.

It is not too much to say that two years ago the whole of this story would have been a maze of bewildering confusion. There are still some points in it that we cannot fully comprehend, for the case is one of far more than ordinary complexity, but the general outlines are now clear. In attempting to elucidate the phenomena it will be remembered that there are no statistics (those given being inapplicable), and the several offspring are only imperfectly referred to the several classes of seeds. This being so, our rationale cannot hope to be complete. Laxton states that as the seeds of peas are liable to change colour with keeping, for this and other reasons he sent to the Society a part of the seeds resulting from his experiment before it was brought to a conclusion.

"The seeds exhibited were derived from a single experiment. Amongst these seeds will be observed some of several remarkable colours, including black, violet, purple-streaked and spotted, maple, grey, greenish, white, and almost every intermediate tint, the varied colours being apparently produced on the outer coat or envelope of the cotyledons only.

* It will be well known to all practical horticulturalists that Laxton, originally of Stamford, made and brought out a large number of the best known modern peas. The firm is now in Bedford.

The peas were selected for their colours, &c., from the third year's sowing in 1869 of the produce of a cross in 1866 of the early round white-seeded and white-flowered garden variety "Ringleader," which is about 2½ ft. in height, fertilised by the pollen of the common purple-flowered "maple" pea, which is taller than "Ringleader," and has slightly indented seeds. I effected impregnation by removing the anthers of the seed-bearer, and applying the pollen at an early stage. This cross produced a pod containing five round white peas, exactly like the ordinary "Ringleader" seeds*.

In 1867 I sowed these seeds, and all five produced tall purple-flowered purplish-stemmed plants†, and the seeds, with few exceptions, had all maple or brownish-streaked envelopes of various shades; the remainder had entirely violet or deep purple-coloured envelopes‡: in shape the peas were partly in-

* A round white ♀ × grey ♂ giving the usual result, round, "white" (yellow) seeds.

† Tall heterozygotes, with normal dominance of purple flowers.

‡ Here we see dominance of the *pigmented* seed-coat as a maternal character over *white* seed-coat. The colours of the seed-coats are described as essentially two: maple or brown-streaked, and violet, the latter being a small minority. As the sequel shows, the latter are heterozygotes, not breeding true. Now Mendel found, and the fact has been confirmed both by Correns and myself, that crossing a grey pea which is capable of producing purple leads to such production as a form of xenia.

We have here therefore in the purple seeds the union of dissimilar gametes, with production of xenia. But as the brown-streaked seeds are also in part heterozygous, the splitting of a compound allelomorph has probably taken place, though without precise statistics and allotment of offspring among the several seeds the point is uncertain. The colour of seed-coats in "grey" peas and probably "maples" also is, as was stated on p. 150, sensitive to conditions, but the whole difference between "maples" and purple is too much to attribute safely to such irregularity. "Maple" is the word used to describe certain seed-coats which are pigmented with intricate brown mottlings on a paler buff ground. In French they are *perdrix*.

dented; but a few were round*. Some of the plants ripened off earlier than the "maple," which, in comparison with "Ringleader," is a late variety; and although the pods were in many instances partially abortive, the produce was very large†.

In 1868 I sowed the peas of the preceding year's growth, and selected various plants for earliness, productiveness, &c. Some of the plants had light-coloured stems and leaves; these all showed white flowers, and produced round white seeds‡. Others had purple flowers, showed the purple on the stems and at the axils of the stipules, and produced seeds with maple, grey, purple-streaked, or mottled, and a few only, again, with violet-coloured envelopes. Some of the seeds were round, some partially indented §. The pods on each plant, in the majority of instances, contained peas of like characters; but in a few cases the peas in the same pod varied slightly, and in some instances a pod or two on the same plant contained seeds all distinct from the remainder‖. The white-flowered plants were generally dwarfish,

* This is not, as it stands, explicable. It seems from this point and also from what follows that if the account is truly given, some of the plants may have been mosaic with segregation of characters in particular flowers; but see subsequent note.

† As, commonly, in heterozygotes when fertile.

‡ Recessive in flower-colour, seed-coat colour, and in seed-shape as a maternal character: pure recessives as the sequel proved.

§ These are then a mixture of pure dominants and cross-bred dominants, and are now inextricably confused. This time the round seeds may have been all on particular plants—showing recessive seed-shape as a maternal character. It seems just possible that this fact suggested the idea of "round" seeds on the *coloured* plants in the last generation. Till that result is confirmed it should be regarded as very doubtful on the evidence. But we cannot at the present time be sure how much difference there was between these *round* seeds and the *normal* maples in point of shape; and on the whole it seems most probable that the roundness was a mere fluctuation, such as commonly occurs among the peas with large indented seeds.

‖ Is this really evidence of segregation of characters, the flower of about the height of "Ringleader"; but the coloured-flowered sorts varied altogether as to height, period of ripening, and colour and shape of seed*. Those seeds with violet-coloured envelopes produced nearly all maple- or parti-coloured seeds, and only here and there one with a violet-coloured envelope; that colour, again, appeared only incidentally, and in a like degree in the produce of the maple-coloured seeds†.

In 1869 the seeds of various selections of the previous year were again sown separately; and the white-seeded peas again produced only plants with white flowers and round white seeds‡. Some of the coloured seeds, which I had expected would produce purple-flowered plants, produced plants with white flowers and round white seeds only§; the majority, however, brought plants with purple flowers and with seeds principally marked with purple or grey, the maple- or brown-streaked being in the minority‖. On some of the purple-flowered plants were again a few pods with peas differing entirely from the remainder on the same plant. In some pods the seeds were all white, in others all black, and in a few, again, all violet¶; but those plants which bore maple-coloured seeds seemed the most constant and fixed in character of the purple-flowered seedlings**, and the purplish and grey peas, being of intermediate characters, ap-

being the unit? In any case the possibility makes the experiment well worth repeating, especially as Correns has seen a phenomenon conceivably similar.

* Being a mixture of heterozygotes (probably involving several pairs of allelomorphs) and homozygotes.

† This looks as if the violet colour was merely due to irregularity of xenia.

‡ Pure recessives.

§ Pure recessives in coats showing maternal dominant character.

‖ Now recognized as pure homozygotes.

¶ This seems almost certainly segregation by flower-units, and is as yet inexplicable on any other hypothesis. Especially paradoxical is the presence of "white" seeds on these plants. The impression is scarcely resistible that some remarkable phenomenon of segregation was really seen here.

** Being now homozygotes.

11—2

peared to vary most*. The violet-coloured seeds again produced almost invariably purplish, grey, or maple peas, the clear violet colour only now and then appearing, either wholly in one pod or on a single pea or two in a pod. All the seeds of the purple-flowered plants were again either round or only partially indented; and the plants varied as to height and earliness. In no case, however, does there seem to have been an intermediate-coloured flower; for although in some flowers I thought I found the purple of a lighter shade, I believe this was owing to light, temperature, or other circumstances, and applied equally to the parent maple. I have never noticed a single tinted white flower nor an indented white seed in either of the three years' produce. The whole produce of the third sowing consisted of seeds of the colours and in the approximate quantities in order as follows,—viz.: 1st, white, about half; 2nd, purplish, grey, and violet (intermediate colours), about three-eighths; and, 3rd, maple about one-eighth.

From the above I gather that the white-flowered white-seeded pea is (if I may use the term) an original variety well fixed and distinct entirely from the maple, that the two do not thoroughly intermingle (for whenever the white flower crops out, the plant and its parts all appear to follow exactly the characters of the white pea), and that the maple is a cross-bred variety which has become somewhat permanent and would seem to include amongst its ancestors one or more bearing seeds either altogether or partly violet- or purple-coloured; for although this colour does not appear on the seed of the "maple," it is very potent in the variety, and appears in many parts of the plant and its offspring from cross-fertilised flowers, sometimes on the external surface or at the sutures of the pods of the latter, at others on the seeds and stems, and very frequently on the seeds; and whenever it shows itself on any part of the plant, the flowers are invariably purple. My deductions have been confirmed by intercrosses effected between the various white-, blue-, some singularly bright green-seeded peas which I have selected, and the maple- and purple-podded and the purple-flowered sugar peas, and by reversing those crosses.

* Being heterozygotes exclusively.

I have also deduced from my experiments, in accordance with the conclusions of the late Mr Knight and others, that the colours of the envelopes of the seeds of peas immediately resulting from a cross are never changed*. I find, however, that the colour and probably the substance of the cotyledons are sometimes, but not always, changed by the cross fertilisation of two different varieties; and I do not agree with Mr Knight that the form and size of the seeds produced are unaltered†; for I have on more than one occasion observed that the cotyledons in the seeds directly resulting from a cross of a blue wrinkled pea fertilised by the pollen of a white round variety have been of a greenish-white colour‡, and the seeds nearly round§ and larger or smaller according as there may have been a difference in the size of the seeds of the two varieties||.

I have also noticed that a cross between a round white and a blue wrinkled pea will in the third and fourth generations (second and third years' produce) at times bring forth blue round, blue wrinkled, white round and white wrinkled peas in the same pods, that the white round seeds, when again sown, will produce only white round seeds, that the white wrinkled seeds will, up to the fourth or fifth generation, produce both blue and white wrinkled and round peas, that the blue round peas will produce blue wrinkled and round peas, but that the blue wrinkled peas will bear only blue wrinkled seeds¶. This

* The nature of this mistake is now clear; for as stated above xenia is only likely to occur when the maternal seed-coat is pigmented. The violet coats in this experiment are themselves cases of xenia.

† Knight, it was seen, crossed round ♀ × indented ♂ and consequently got no change of form.

‡ Cotyledons seen through coat.

§ Ordinary dominance of round.

|| This is an extraordinary statement to be given as a general truth. There are sometimes indications of this kind, but certainly the facts are not usually as here stated.

¶ If we were obliged to suppose that this is a matured conclusion based on detailed observation it would of course constitute the most serious "exception" yet recorded. But it is clear that the five would seem to indicate that the white round and the blue wrinkled peas are distinct varieties derived from ancestors respectively possessing one only of those marked qualities; and, in my opinion, the white round peas trace their origin to a dwarfish pea having white flowers and round white seeds, and the blue wrinkled varieties to a tall variety, having also white flowers but blue wrinkled seeds. It is also noticeable, that from a single cross between two different peas many hundreds of varieties, not only like one or both parents and intermediate, but apparently differing from either, may be produced in the

statements are not mutually consistent. We have dominance of round white in first cross.

In the second generation blue wrinkled give only blue wrinkled, and blue round give blue wrinkled and round, in accordance with general experience. But we are told that white round give *only* white round. This would be true of some white rounds, but not, according to general experience, of all. Lastly we are told *white wrinkled give all four classes.* If we had not been just told by Laxton that the first cross showed dominance of white round, and that blue wrinkled and blue round give the Mendelian result, I should hesitate in face of this positive statement, but as it is inconsistent with the rest of the story I think it is unquestionably an error of statement. The context, and the argument based on the maple crosses show clearly also what was in Laxton's mind. He plainly expected the characters of the original pure varieties to separate out according to their original combinations, and this expectation confused his memory and general impressions. This, at least, until any such result is got by a fresh observer, using strict methods, is the only acceptable account.

Of the same nature is the statement given by the late Mr Masters to Darwin (*Animals and Plants*, I. p. 318) that blue round, white round, blue wrinkled, and white wrinkled, all reproduced all four sorts during successive years. Seeing that one sort would give all four, and two would give two kinds, without special counting such an impression might easily be produced. There are the further difficulties due to seed-coat colour, and the fact that the distinction between round and wrinkled may need some discrimination. The sorts are not named, and the case cannot be further tested.

course of three or four years (the shortest time which I have ascertained it takes to attain the climax of variation in the produce of cross-fertilised peas, and until which time it would seem useless to expect a fixed seedling variety to be produced*), although a reversion to the characters of either parent, or of any one of the ancestors, may take place at an earlier period.

These circumstances do not appear to have been known to Mr Knight, as he seems to have carried on his experiments by continuing to cross his seedlings in the year succeeding their production from a cross and treating the results as reliable; whereas it is probable that the results might have been materially affected by the disturbing causes then in existence arising from the previous cross fertilisation, and which, I consider, would, in all cases where either parent has not become fixed or permanent, lead to results positively perplexing and uncertain, and to variations almost innumerable. I have again selected, and intend to sow, watch, and report; but as the usual climax of variation is nearly reached in the recorded experiment, I do not anticipate much further deviation, except in height and period of ripening—characters which are always very unstable in the pea. There are also important botanical and other variations and changes occurring in cross-fertilised peas to which it is not my province here to allude; but in conclusion I may, perhaps, in furtherance of the objects of this paper, be permitted to inquire whether any light can, from these observations or other means, be thrown upon the origin of the cultivated kinds of peas, especially the "maple" variety, and also as to the source whence the violet and other colours which appear at intervals on the seeds and in the offspring of cross-fertilised purple-flowered peas are derived."

The reader who has closely followed the preceding passage will begin to appreciate the way in which the new principles help us to interpret these hitherto paradoxical phenomena. Even in this case, imperfectly recorded as it is, we can form a fairly clear idea of what was taking place.

* See later.

If the "round" seeds really occurred as a distinct class, on the heterozygotes as described, it is just possible that the fact may be of great use hereafter.

We are still far from understanding maternal seed-form—and perhaps size—as a dominant character. So far, as Miss Saunders has pointed out to me, it appears to be correlated with a thick and coloured seed-coat.

We have now seen the nature of Professor Weldon's collection of contradictory evidence concerning dominance in peas. He tells us: "Enough has been said to show the grave discrepancy between the evidence afforded by Mendel's experiments and that obtained by observers equally trustworthy."

He proceeds to a discussion of the *Telephone* and *Telegraph* group and recites facts, which I do not doubt for a moment, showing that in this group of peas—which have unquestionably been more or less "blend" or "mosaic" forms from their beginning—the "laws of dominance and segregation" do not hold. Professor Weldon's collection of the facts relating to *Telephone*, &c. has distinct value, and it is the chief addition he makes to our knowledge of these phenomena. The merit however of this addition is diminished by the erroneous conclusion drawn from it, as will be shown hereafter. Meanwhile the reader who has studied what has been written above on the general questions of stability, "purity," and "universal" dominance, will easily be able to estimate the significance of these phenomena and their applicability to Mendel's hypotheses.

D. Miscellaneous cases in other plants and animals.

Professor Weldon proceeds :

"In order to emphasize the need that the ancestry of the parents, used in crossing, should be considered in discussing the results of a cross, it may be well to give one or two more examples of fundamental inconsistency between different competent observers."

The "one or two" run to three, viz. Stocks (hoariness and colour); *Datura* (character of fruits and colour of flowers); and lastly colours of Rats and Mice. Each of these subjects, as it happens, has been referred to in the forthcoming paper by Miss Saunders and myself. *Datura* and *Matthiola* have been subjected to several years' experiment and I venture to refer the reader who desires to see whether the facts are or are not in accord with Mendel's expectation and how far there is "fundamental inconsistency" amongst them to a perusal of our work.

But as Professor Weldon refers to some points that have not been explicitly dealt with there, it will be safer to make each clear as we proceed.

1. *Stocks* (*Matthiola*). Professor Weldon quotes Correns' observation that glabrous Stocks crossed with hoary gave offspring all hoary, while Trevor Clarke thus obtained some hoary and some glabrous. As there are some twenty different sorts of Stocks* it is not surprising that different observers should have chanced on different materials and obtained different results. Miss Saunders

* The number in Haage and Schmidt's list exceeds 200, counting colour-varieties.

has investigated laws of heredity in Stocks on a large scale and an account of her results is included in our forthcoming Report. Here it must suffice to say that the cross hoary ♀ × glabrous ♂ always gave offspring all hoary except once: that the cross glabrous ♀ × hoary ♂ of several types gave all hoary; *but* the same cross using other hoary types did frequently give a mixture, some of the offspring being hoary, others glabrous. Professor Weldon might immediately decide that here was the hoped for phenomenon of "reversed" dominance, due to ancestry, but here again that hypothesis is excluded. For the glabrous (recessive) cross-breds were *pure*, and produced on self-fertilisation glabrous plants only, being in fact, almost beyond question, "false hybrids" (see p. 34), a specific phenomenon which has nothing to do with the question of dominance.

Professor Weldon next suggests that there is discrepancy between the observations as to flower-colour. He tells us that Correns found *violet* Stocks crossed with "*yellowish white*" gave violet or shades of violet flaked together. According to Professor Weldon

> "On the other hand Nobbe crossed a number of varieties of *M. annua* in which the flowers were white, violet, carmine-coloured, crimson or dark blue. These were crossed in various ways, and before a cross was made the colour of each parent was matched by a mixture of dry powdered colours which was preserved. In every case the hybrid flower was of an intermediate colour, which could be matched by mixing the powders which recorded the parental colours. The proportions in which the powders were mixed are not given in each [any] case, but it is clear that the colours blended*."

* The original passage is in *Landwirths. Versuchstationen*, 1888, XXXV. [*not* XXXIV.], p. 151.

On comparing Professor Weldon's version with the originals we find the missing explanations. Having served some apprenticeship to the breeding of Stocks, we, here, are perhaps in a better position to take the points, but it is to me perfectly inexplicable how in such a simple matter as this he can have gone wrong.

Note then

(1) That Nobbe does *not* specify *which* colours he crossed together, beyond the fact that *white* was crossed with each fertile form. The *crimson* form (*Karmoisinfarbe*), being double to the point of sterility, was not used. There remain then, white, carmine, and two purples (violet, "dark blue"). When *white* was crossed with either of these, Nobbe says the colour becomes *paler*, whichever sort gave the pollen. Nobbe does not state that he crossed *carmine* with the purples.

(2) Professor Weldon gives no qualification in his version. Nobbe however states that he found it very difficult to distinguish the result of crossing *carmine with white* from that obtained by crossing *dark blue or violet with white**, thereby nullifying Professor Weldon's statement that in every case the cross was a simple mixture of the parental colours—a proposition sufficiently disproved by Miss Saunders' elaborate experiments.

(3) Lately the champion of the "importance of small variations," Professor Weldon now prefers to treat the distinctions between established varieties as negligible

* *"Es ist sogar sehr schwierig, einen Unterschied in der Farbe der Kreuzungsprodukte von Karmin und Weiss gegenüber Dunkelblau oder Violett und Weiss zu erkennen."*

fluctuations instead of specific phenomena*. Therefore when Correns using "*yellowish white*" obtained one result and Nobbe using "*white*" obtained another, Professor Weldon hurries to the conclusion that the results are comparable and therefore contradictory. Correns however though calling his flowers *gelblich-weiss* is careful to state that they are described by Haage and Schmidt (the seedmen) as "*schwefel-gelb*" or sulphur-yellow. The topics Professor Weldon treats are so numerous that we cannot fairly expect him to be personally acquainted with all; still had he *looked* at Stocks before writing, or even at the literature relating to them, he would have easily seen that these yellow Stocks are a thoroughly distinct form†; and in accordance with this fact it would be surprising if they had not a distinctive behaviour in their crosses. To use our own terminology their colour character depends almost certainly on a *compound* allelomorph. Consequently there is no evidence of contradiction in the results, and appeal to ancestry is as unnecessary as futile.

2. *Datura*. As for the evidence on *Datura*, I must refer the reader again to the experiments set forth in our Report.

The phenomena obey the ordinary Mendelian rules with accuracy. There are (as almost always where discontinuous

* See also the case of *Buchsbaum*, p. 146, which received similar treatment.

† One of the peculiarities of most *double* "sulphur" races is that the singles they throw are *white*. See Vilmorin, *Fleurs de pleine Terre*, 1866, p. 354, *note*. In *Wien. Ill. Gartenztg.* 1891, p. 74, mention is made of a new race with singles also "sulphur," cp. *Gartenztg.* 1884, p. 46. Messrs Haage and Schmidt have kindly written to me that this new race has the alleged property, but that six other yellow races (two distinct colours) throw their singles white.

variation is concerned) occasional cases of "mosaics," a phenomenon which has nothing to do with "ancestry."

3. *Colours of Rats and Mice.* Professor Weldon reserves his collection of evidence on this subject for the last. In it we reach an indisputable contribution to the discussion—a reference to Crampe's papers, which together constitute without doubt the best evidence yet published, respecting colour-heredity in an animal. So far as I have discovered, the only previous reference to these memoirs is that of Ritzema Bos*, who alludes to them in a consideration of the alleged deterioration due to in-breeding.

Now Crampe through a long period of years made an exhaustive study of the peculiarities of the colour-forms of Rats, white, black, grey and their piebalds, as exhibited in Heredity.

Till the appearance of Professor Weldon's article Crampe's work was unknown to me, and all students of Heredity owe him a debt for putting it into general circulation. My attention had however been called by Dr Correns to the interesting results obtained by von Guaita, experimenting with crosses originally made between albino *mice* and piebald Japanese waltzing mice. This paper also gives full details of an elaborate investigation admirably carried out and recorded.

In the light of modern knowledge both these two researches furnish material of the most convincing character demonstrating the Mendelian principles. It would be a useful task to go over the evidence they contain and rearrange it in illustration of the laws now perceived. To do this here is manifestly impossible, and it must suffice to point out that the albino is a simple recessive in both cases (the

* *Biol. Cblt.* XIV. 1894, p. 79.

waltzing character in mice being also a recessive), and that the "wild grey" form is one of the commonest heterozygotes—there appearing, like the yellow cotyledon-colour of peas, *in either of two capacities,* i.e. as a pure form, or as the heterozygote form of one or more combinations*.

Professor Weldon refers to both Crampe and von Guaita, whose results show an essential harmony in the fact that both found *albino* an obvious recessive, pure almost without exception, while the coloured forms show various phenomena of dominance. Both found heterozygous colour-types. He then searches for something that looks like a contradiction. Of this there is no lack in the works of Johann von Fischer (11)—an authority of a very different character—whom he quotes in the following few words:

"In both rats and mice von Fischer says that piebald rats crossed with albino varieties of their species, give piebald young if the father only is piebald, white young if the mother only is piebald."

But this is doing small justice to the completeness of Johann von Fischer's statement, which is indeed a proposition of much more amazing import.

That investigator in fact began by a study of the cross between the albino Ferret and the Polecat, as a means of testing whether they were two species or merely varieties. The cross, he found, was in colour and form a blend of the parental types. Therefore, he declares, the Ferret and the

* The various "contradictions" which Professor Weldon suggests exist between Crampe, von Guaita and Colladon can almost certainly be explained by this circumstance. For Professor Weldon "wild-coloured" mice, however produced, are "wild-coloured" mice and no more (see Introduction).

Polecat are two distinct species, because, "as everybody ought to know,"

"*The result of a cross between albino and normal [of one species] is always a constant one, namely an offspring like the father at least in colour**,"

whereas in *crosses* (between species) this is *not* the case.

And again, after reciting that the Ferret-Polecat crosses gave intermediates, he states:

"But all this is *not* the case in crosses between albinos and normal animals within the species, in which always and without any exception the young resemble the father in colour†."

These are admirable illustrations of what is meant by a "*universal*" proposition. But von Fischer doesn't stop here. He proceeds to give a collection of evidence in proof of this truth which he says "ought to be known to everyone." He has observed the fact in regard to albino mole, albino shrew (*Sorex araneus*), melanic squirrel (*Sciurus vulgaris*), albino ground-squirrel (*Hypudaeus terrestris*), albino hamster, albino rats, albino mice, piebald (grey-and-white or black-and-white) mice and rats, partially albino sparrow, and we are even presented with two cases in Man. No single exception was known to von Fischer‡.

* "Das Resultat einer Kreuzung zwischen Albino- und Normalform ist stets, also, constant, ein dem Vater mindestens in der Färbung gleiches Junge." This law is predicated for the case in which both parents belong to the same species.

† "Dieses Alles ist aber *nie* der Fall bei Kreuzungen unter Leucismen und normalen Thieren innerhalb der Species, bei denen *stets und ohne jede Ausnahme die Jungen in Färbung dem Vater gleichen.*"

‡ He even withdraws two cases of his own previously published, in which grey and albino mice were alleged to have given mixtures, saying that this result must have been due to the broods having been accidentally mixed by the servants in his absence.

In his subsequent paper von Fischer declares that from matings of rats in which the mothers were grey and the fathers albino he bred 2017 pure albinos; and from albino mothers and grey fathers 3830 normal greys. "Not a single individual varied in any respect, or was in any way intermediate."

With piebalds the same result is asserted, save that certain melanic forms appeared. Finally von Fischer repeats his laws already reached, giving them now in this form: *that if the offspring of a cross show only the colour of the father, then the parents are varieties of one species; but if the colour of the offspring be intermediate or different from that of the father, then the parents belong to distinct species.*

The reader may have already gathered that we have here that bane of the advocate—the witness who proves too much. But why does Professor Weldon confine von Fischer to the few modest words recited above? That author has—so far as colour is concerned—a complete law of heredity supported by copious "observations." Why go further?

Professor Weldon "brings forth these strong reasons" of the rats and mice with the introductory sentence:

> "Examples might easily be multiplied, but as before, I have chosen rather to cite a few cases which rest on excellent authority, than to quote examples which may be doubted. I would only add one case among animals, in which the evidence concerning the inheritance of colour is affected by the ancestry of the varieties used."

So once again Professor Weldon suggests that his laws of ancestry will explain even the discrepancies between von Fischer on the one hand and Crampe and von Guaita on the other but he does not tell us how he proposes to apply them.

In the cross between the albino and the grey von Fischer tells us that both colours appear in the offspring, but always, without exception or variation, that of the father only, in 5847 individuals.

Surely, the law of ancestry, if he had a moment's confidence in it, might rather have warned Professor Weldon that von Fischer's results were wrong somewhere, of which there cannot be any serious doubt. The precise source of error is not easy to specify, but probably carelessness and strong preconception of the expected result were largely responsible, though von Fischer says he did all the recording most carefully himself.

Such then is the evidence resting "on excellent authority": may we some day be privileged to see the "examples which may be doubted"?

The case of mice, invoked by Professor Weldon, has also been referred to in our Report. Its extraordinary value as illustrating Mendel's principles and the beautiful way in which that case may lead on to extensions of those principles are also there set forth (see the present Introduction, p. 25). Most if not all of such "conflicting" evidence can be reconciled by the steady application of the Mendelian principle that the progeny will be constant when—and only when*—*similar* gametes meet in fertilisation, apart from any question of the characters of the parent which produces those gametes.

* Excluding "false hybridisations."

V. PROFESSOR WELDON'S QUOTATIONS FROM LAXTON.

In support of his conclusions Professor Weldon adduces two passages from Laxton, some of whose testimony we have just considered. This further evidence of Laxton is so important that I reproduce it in full. The first passage, published in 1866, is as follows:—

"The results of experiments in crossing the Pea tend to show that the colour of the immediate offspring or second generation sometimes follows that of the female parent, is sometimes intermediate between that and the male parent, and is sometimes distinct from both; and although at times it partakes of the colour of the male, it has not been ascertained by the experimenter ever to follow the exact colour of the male parent*. In shape, the seed frequently has an intermediate character, but as often follows that of either parent. In the second generation, in a single pod, the result of a cross of Peas different in shape and colour, the seeds are sometimes all intermediate, sometimes represent either or both parents in shape or colour, and sometimes both colours and characters, with their intermediates, appear. The results also seem to show that the third generation or the immediate offspring of a cross, frequently varies from its parents in a limited manner—usually in one direction only, but that the fourth generation produces numerous and wider variations†; the seed often reverting partly to the colour and character of its ancestors of the first generation, partly partaking of the various intermediate colours and characters, and partly sporting quite away from any of its ancestry."

* This is of course on account of the maternal seed characters. Unless the coat-characters are treated separately from the cotyledon-characters Laxton's description is very accurate. Both this and the statements respecting the "shape" of the seeds, a term which as used by Laxton means much more than merely "wrinkled" and "smooth," are recognizably true as general statements.

† Separation of hypallelomorphs.

Here Professor Weldon's quotation ceases. It is unfortunate he did not read on into the very next sentence with which the paragraph concludes :—

"These sports appear to become fixed and permanent in the next and succeeding generations; and the tendency to revert and sport thenceforth seems to become checked if not absolutely stopped*."

Now if Professor Weldon instead of leaving off on the word "ancestry" had noticed this passage, I think his article would never have been written.

Laxton proceeds :—

"The experiments also tend to show that the height of the plant is singularly influenced by crossing; a cross between two dwarf peas, commonly producing some dwarf and some tall [? in the second generation]; but on the other hand, a cross between two tall peas does not exhibit a tendency to diminution in height.

"No perceptible difference appears to result from reversing the parents; the influence of the pollen of each parent at the climax or fourth generation producing similar results†."

The significance of this latter testimony I will presently discuss.

Professor Weldon next appeals to a later paper of Laxton's published in 1890. From it he quotes this passage :

"By means, however, of cross-fertilisation alone, and unless it be followed by careful and continuous selection, the labours of the cross-breeder, instead of benefiting the gardener, may lead to utter confusion,"

* The combinations being exhausted. Perhaps Professor Weldon thought his authority was here lapsing into palpable nonsense!

† Laxton constantly refers to this conception of the "climax" of—as we now perceive—analytical variation and recombination. Many citations could be given respecting his views on this "climax" (cp. p. 167).

Here again the reader would have gained had Professor Weldon, instead of leaving off at the comma, gone on to the end of the paragraph, which proceeds thus:—

"because, as I have previously stated, the Pea under ordinary conditions is much given to sporting and reversion, for when two dissimilar old or fixed varieties have been cross-fertilised, three or four generations at least must, under the most favourable circumstances, elapse before the progeny will become fixed or settled; and from one such cross I have no doubt that, by sowing every individual Pea produced during the three or four generations, hundreds of different varieties may be obtained; but as might be expected, I have found that where the two varieties desired to be intercrossed are unfixed, confusion will become confounded*, and the variations continue through many generations, the number at length being utterly incalculable."

Professor Weldon declares that Laxton's "experience was altogether different from that of Mendel." The reader will bear in mind that when Laxton speaks of fixing a variety he is not thinking particularly of seed-characters, but of all the complex characters, fertility, size, flavour, season of maturity, hardiness, etc., which go to make a serviceable pea. Considered carefully, Laxton's testimony is so closely in accord with Mendelian expectation that I can imagine no chance description in non-Mendelian language more accurately stating the phenomena.

Here we are told in unmistakable terms the breaking up of the original combination of characters on crossing, their re-arrangement, that at the fourth or fifth generation the possibilities of sporting [sub-division of compound allelomorphs and re-combinations of them ?] are exhausted, that there are then definite forms which if selected are

* Further subdivision and recombination of hypallelomorphs.

thenceforth fixed [produced by union of similar gametes?] that it takes longer to select some forms [dominants?] than others [recessives?], that there may be "mule" forms* or forms which cannot be fixed at all† [produced by union of dissimilar gametes?].

But Laxton tells us more than this. He shows us that numbers of varieties may be obtained—hundreds—"incalculable numbers." Here too if Professor Weldon had followed Mendel with even moderate care he would have found the secret. For in dealing with the crosses of *Phaseolus* Mendel clearly forecasts the conception of *compound characters themselves again consisting of definite units*, all of which may be separated and re-combined in the possible combinations, laying for us the foundation of the new science of Analytical Biology.

How did Professor Weldon, after reading Mendel, fail to perceive these principles permeating Laxton's facts? Laxton must have seen the very things that Mendel saw, and had he with his other gifts combined that penetration which detects a great principle hidden in the thin mist of "exceptions," we should have been able to claim for him that honour which must ever be Mendel's in the history of discovery.

When Laxton speaks of selection and the need for it, he means, what the raiser of new varieties almost always means, the selection of *definite* forms, not impalpable fluctuations. When he says that without selection there will be utter confusion, he means—to use Mendelian terms

* For instance the *talls* produced by crossing *dwarfs* are such "mules." Tschermak found in certain cases distinct increase in height in such a case, though not always (p. 531).

† "The remarkably fine but unfixable pea *Evolution*." Laxton, p. 37.

—that the plant which shows the desired combination of characters must be chosen and bred from, and that if this be not done the grower will have endless combinations mixed together in his stock. If however such a selection be made in the fourth or fifth generation the breeder may very possibly have got a fixed form—namely, one that will breed true*. On the other hand he may light on one that does not breed true, and in the latter case it may be that the particular type he has chosen is not represented in the gametes and will *never* breed true, though selected to the end of time. Of all this Mendel has given us the simple and final account.

At Messrs Sutton and Sons, to whom I am most grateful for unlimited opportunities of study, I have seen exactly such a case as this. For many years Messrs Sutton have been engaged in developing new strains of the Chinese Primrose (*Primula sinensis*, hort.). Some thirty thoroughly distinct and striking varieties (not counting the *Stellata* or "Star" section) have already been produced which breed true or very nearly so. In 1899 Messrs Sutton called my attention to a strain known as "Giant Lavender," a particularly fine form with pale magenta or lavender flowers, telling me that it had never become fixed. On examination it appeared that self-fertilised seed saved from this variety gave some magenta-reds, some lavenders, and some which are white on opening but tinge with very faint pink as the flower matures.

On counting these three forms in two successive years the following figures appeared. Two separately bred batches raised from "Giant Lavender" were counted in each year.

* Apart from fresh original variations, and perhaps in some cases imperfect homozygosis of some hypallelomorphs.

	Magenta red	Lavender	White faintly tinged
1901 1st batch	19	27	14
„ 2nd „	9	20	9
1902 1st „	12	23	11
„ 2nd „	14	26	11
	—	—	—
	54	96	45

The numbers 54 : 96 : 45 approach the ratio 1 : 2 : 1 so nearly that there can be no doubt we have here a simple case of Mendelian laws, operating without definite dominance, but rather with blending.

When Laxton speaks of the "remarkably fine but unfixable pea *Evolution*" we now know for the first time exactly what the phenomenon meant. It, like the "Giant Lavender," was a "mule" form, not represented by germ-cells, and in each year arose by "self-crossing."

This is only one case among many similar ones seen in the Chinese Primrose. In others there is no doubt that more complex factors are at work, the subdivision of compound characters, and so on. The history of the "Giant Lavender" goes back many years and is not known with sufficient precision for our purposes, but like all these forms it originated from crossings among the old simple colour varieties of *sinensis.*

VI. The Argument Built on Exceptions.

So much for the enormous advance that the Mendelian principles already permit us to make. But what does Professor Weldon offer to substitute for all this? Nothing.

Professor Weldon suggests that a study of ancestry will help us. Having recited Tschermak's exceptions and the great irregularities seen in the *Telephone* group, he writes:

"Taking these results together with Laxton's statements, and with the evidence afforded by the *Telephone* group of hybrids, I think we can only conclude that segregation of seed-characters is not of universal occurrence among cross-bred peas, and that when it does occur, it may or may not follow Mendel's law."

Premising that when pure types are used the exceptions form but a small part of the whole, and that any supposed absence of "segregation" may have been *variation*, this statement is perfectly sound. He proceeds:—

"The law of segregation, like the law of dominance, appears therefore to hold only for races of *particular ancestry* [my italics]. In special cases, other formulae expressing segregation have been offered, especially by De Vries and by Tschermak for other plants, but these seem as little likely to prove generally valid as Mendel's formula itself.

"The fundamental mistake which vitiates all work based upon Mendel's method is the neglect of ancestry, and the attempt to regard the whole effect upon offspring, produced by a particular parent, as due to the existence in the parent of particular structural characters; while the contradictory results obtained by those who have observed the offspring of parents identical in certain characters show clearly enough that not only the parents themselves, but their race, that is their ancestry, must be taken into account before the result of pairing them can be predicted."

In this passage the Mendelian view is none too precisely represented. I should rather have said that it was from Mendel, first of all men, that we have learnt *not* to regard the effects produced on offspring "as due to the existence in the parent of particular structural characters." We have come rather to disregard the particular structure of the parent except in so far as it may give us a guide as to the nature of its gametes.

This indication, if taken in the positive sense—as was sufficiently shown in considering the significance of the "mule" form or "hybrid-character"—we now know may be absolutely worthless, and in any unfamiliar case is very likely to be so. Mendel has proved that the inheritance from individuals of *identical ancestry* may be entirely different: that from identical ancestry, without new variation, may be produced three kinds of individuals (in respect of each pair of characters), namely, individuals capable of transmitting one type, or another type, or both: moreover that the statistical relations of these three classes of individuals to each other will in a great number of cases be a definite one: and of all this he shows a complete account.

Professor Weldon cannot deal with any part of this phenomenon. He does little more than allude to it in passing and point out exceptional cases. These he suggests a study of ancestry will explain.

As a matter of fact a study of ancestry will give little guide—perhaps none—even as to the probability of the phenomenon of dominance of a character, none as to the probability of normal "purity" of germ-cells. Still less will it help to account for fluctuations in dominance, or irregularities in "purity."

Ancestry and Dominance.

In a series of astonishing paragraphs (pp. 241–2) Professor Weldon rises by gradual steps, from the exceptional facts regarding occasional dominance of green colour in *Telephone* to suggest that the *whole phenomenon of dominance may be attributable to ancestry*, and that in fact one character has no natural dominance over another, apart from what has been created by selection of ancestry. This piece of reasoning, one of the most remarkable examples of special pleading to be met with in scientific literature, must be read as a whole. I reproduce it entire, that the reader may appreciate this curious effort. The remarks between round parenthetical marks are Professor Weldon's, those between crotchets are mine.

"Mendel treats such characters as yellowness of cotyledons and the like as if the condition of the character in two given parents determined its condition in all their subsequent offspring*. Now it is well known to breeders, and is clearly shown in a number of cases by Galton and Pearson, that the condition of an animal does not as a rule depend upon the condition of any one pair of ancestors alone, but in varying degrees upon the condition of all its ancestors in every past generation, the condition in each of the half-dozen nearest generations having a quite sensible effect. Mendel does not take the effect of differences of ancestry into account, but considers that any yellow-seeded pea, crossed with any green-seeded pea, will behave in a certain definite way, whatever the ancestry of the green and yellow peas may have been. (He does not say this in words, but his attempt to treat his results as generally true of the characters observed is unintelligible unless this hypothesis be assumed.) The experiments afford no evidence which can be held to justify this hypothesis. His observations on cotyledon colour, for example, are based upon 58 cross-fertilised flowers, all of which were borne upon ten plants; and we are not even told whether these ten plants included individuals from more than two races.

"The many thousands of individuals raised from these ten

* Mendel, on the contrary, disregards the "condition of the character" in the parent altogether; but is solely concerned with the nature of the characters of the *gametes*.

plants afford an admirable illustration of the effect produced by crossing a few pairs of plants of known ancestry; but while they show this perhaps better than any similar experiment, they do not afford the data necessary for a statement as to the behaviour of yellow-seeded peas in general, whatever their ancestry, when crossed with green-seeded peas of any ancestry. [Mendel of course makes no such statement.]

"When this is remembered, the importance of the exceptions to dominance of yellow cotyledon-colour, or of smooth and rounded shape of seeds, observed by Tschermak, is much increased; because although they form a small percentage of his whole result, they form a very large percentage of the results obtained with peas of certain races. [Certainly.] The fact that *Telephone* behaved in crossing on the whole like a green-seeded race of exceptional dominance shows that something other than the mere character of the parental generation operated in this case. Thus in eight out of 27 seeds from the yellow *Pois d'Auvergne* ♀ × *Telephone* ♂ the cotyledons were yellow with green patches; the reciprocal cross gave two green and one yellow-and-green seed out of the whole ten obtained; and the cross *Telephone* ♀ × (yellow-seeded) *Buchsbaum** ♂ gave on one occasion two green and four yellow seeds.

"So the cross *Couturier* (orange-yellow) ♀ × the green-seeded *Express* ♂ gave a number of seeds intermediate in colour. (It is not clear from Tschermak's paper whether *all* the seeds were of this colour, but certainly some of them were.) The green *Plein le Panier* [*Fillbasket*] ♀ × *Couturier* ♂ in three crosses always gave either seeds of colour intermediate between green and yellow, or some yellow and some green seeds in the same pod. The cross reciprocal to this was not made; but *Express* ♀ × *Couturier* ♂ gave 22 seeds of which four were yellowish green†.

"These facts show *first* that Mendel's law of dominance conspicuously fails for crosses between certain races, while it

* Regarding this "exception" see p. 146.

† See p. 148.

appears to hold for others; and *secondly* that the intensity of a character in one generation of a race is no trustworthy measure of its dominance in hybrids. The obvious suggestion is that the behaviour of an individual when crossed depends largely upon the characters of its ancestors*. When it is remembered that peas are normally self-fertilised, and that more than one named variety may be selected out of the seeds of a single hybrid pod, it is seen to be probable that Mendel worked with a very definite combination of ancestral characters, and had no proper basis for generalisation about yellow and green peas of any ancestry" [which he never made].

Let us pause a moment before proceeding to the climax. Let the reader note we have been told of *two* groups of cases in which dominance of yellow failed or was irregular. (Why are not Gärtner's and Seton's "exceptions" referred to here?) In one of these groups *Couturier* was always one parent, either father or mother, and were it not for Tschermak's own obvious hesitation in regard to his own exceptions (see p. 148), I would gladly believe that *Couturier*—a form I do not know—may be an exceptional variety. *How* Professor Weldon proposes to explain its peculiarities by reference to ancestry he omits to tell us. The *Buchsbaum* case is already disposed of, for on Tschermak's showing, it is an unstable form.

Happily, thanks to Professor Weldon, we know rather more of the third case, that of *Telephone,* which, whether as father or mother, was frequently found by Tschermak to give either green, greenish, or patchwork-seeds when crossed with yellow varieties. It behaves, in short, "like a green-seeded pea of exceptional dominance," as we are now told. For this dominant quality of *Telephone's* greenness we are asked to account *by appeal to its ancestry.* May we not

* Where was that "logician," the "consulting-partner," when this piece of reasoning passed the firm?

expect, then, this *Telephone* to be—if not a pure-bred green pea from time immemorial—at least as pure-bred as other green peas which do *not* exhibit dominance of green at all? Now, what is *Telephone*? Do not let us ask too much. Ancestry takes a lot of proving. We would not reject him "*parce qu'il n'avait que soixante & onze quartiers, & que le reste de son arbre généalogique avait été perdu par l'injure du tems.*"

But with stupefaction we learn from Professor Weldon himself that *Telephone* is the very variety which he takes *as his type of a permanent and incorrigible mongrel*, a character it thoroughly deserves.

From *Telephone* he made his colour scale! Tschermak declares the cotyledons to be "yellowish or whitish green, often entirely bright yellow*." So little is it a thoroughbred green pea, that it cannot always keep its own self-fertilised offspring green. Not only is this pea a parti-coloured mongrel, but Professor Weldon himself quotes Culverwell that as late as 1882 both *Telegraph* and *Telephone* "will always come from one sort, more especially from the green variety"; and again regarding a supposed good sample of *Telegraph* that "Strange to say, although the peas were taken from one lot, those sown in January produced a great proportion of the light variety known as *Telephone*. These were of every shade of light green up to white, and could have been shown for either variety," *Gard. Chron.* 1882 (2), p. 150. This is the variety whose green, it is suggested, partially "dominates" over the yellow of *Pois d'Auvergne*, a yellow variety which has a clear lineage of about a century, and probably more. If, therefore, the facts regarding *Telephone* have any bearing on the signi-

* "*Speichergewebe gelblich—oder weisslich—grün, manchmal auch vollständig hellgelb.*" Tschermak (36), p. 480.

ficance of ancestry, they point the opposite way from that in which Professor Weldon desires to proceed.

In view of the evidence, the conclusion is forced upon me that the suggestion that "ancestry" may explain the facts regarding *Telephone* has no meaning behind it, but is merely a verbal obstacle. Two words more on *Telephone*. On p. 147 I ventured to hint that if we try to understand the nature of the appearance of green in the offspring of *Telephone* bred with yellow varieties, we are more likely to do so by comparing the facts with those of false hybridisation than with fluctuations in dominance. In this connection I would call the reader's attention to a point Professor Weldon misses, that Tschermak *also got yellowish-green seeds from Fillbasket* (*green*) *crossed with Telephone*. I suggest therefore that *Telephone's* allelomorphs may be in part transmitted to its offspring in a state which needs no union with any corresponding allelomorph of the other gamete, just as may the allelomorphs of "false hybrids." It would be quite out of place here to pursue this reasoning, but the reader acquainted with special phenomena of heredity will probably be able fruitfully to extend it. It will be remembered that we have already seen the further fact that the behaviour of *Telephone* in respect to seed-shape was also peculiar (see p. 152).

Whatever the future may decide on this interesting question it is evident that with *Telephone* (and possibly *Buchsbaum*) we are encountering a *specific* phenomenon, which calls for specific elucidation and not a case simply comparable with or contradicting the evidence of dominance in general.

In this excursion we have seen something more of the "exceptions." Many have fallen, but some still stand, though even as to part of the remainder Tschermak entertains some doubts, and, it will be remembered, cautions his reader that of his exceptions some may be self-fertilisations, and some did not germinate*. Truly a slender basis to carry the coming structure!

But Professor Weldon cannot be warned. He told us the "law of dominance conspicuously fails for crosses between certain races." Thence the start. I venture to give the steps in this impetuous argument. There are exceptions †—a fair number if we count the bad ones—there may be more—must be more—*are* more—no doubt many more: so to the brink. Then the bold leap: may there not be as many cases one way as the other? We have not tried half the sorts of Peas yet. There is still hope. True we know dominance of many characters in some hundreds of crosses, using some twenty varieties—not to speak of other plants and animals—but we *do* know some exceptions, of which a few are still good. So dominance

* In his latest publication on this subject, the notes to the edition of Mendel in Ostwald's *Klassiker* (pp. 60—61), Tschermak, who has seen more true exceptions than any other observer, thus refers to them. As to dominance:—"*Immerhin kommen vereinzelt auch zweifellose Fälle von Merkmalmischung, d. h. Uebergangsformen zwischen gelber und grüner Farbe, runder und runzeliger Form vor, die sich in weiteren Generationen wie dominantmerkmalige Mischlinge verhalten.*" As to purity of the extracted recessives:—*Ganz vereinzelt scheinen Ausnahmsfälle vorzukommen.*"

Küster (22) also in a recent note on Mendelism points out, with reason, that the number of "exceptions" to dominance that we shall find, depends simply on the stringency with which the supposed "law" is drawn. The same writer remarks further that Mendel makes no such rigid definition of dominance as his followers have done.

† If the "logician-consulting-partner" will successfully apply this *Fallacia acervalis*, the "method of the vanishing heap," to dominant peas, he will need considerable leisure.

may yet be all a myth, built up out of the petty facts those purblind experimenters chanced to gather. Let us take wider views. Let us look at fields more propitious—more what we would have them be! Let us turn to eye-colour: at least there is no dominance in that. Thus Professor Weldon, telling us that Mendel "had no proper basis for generalisation about yellow and green peas of any ancestry," proceeds to this lamentable passage:—

"Now in such a case of alternative inheritance as that of human eye-colour, it has been shown that a number of pairs of parents, one of whom has dark and the other blue eyes, will produce offspring of which nearly one half are dark-eyed, nearly one half are blue-eyed, a small but sensible percentage being children with mosaic eyes, the iris being a patch-work of lighter and darker portions. But the dark-eyed and light-eyed children are not equally distributed among all families; and it would almost certainly be possible, by selecting cases of marriage between men and women of appropriate ancestry, to demonstrate for their families a law of dominance of dark over light eye-colour, or of light over dark. Such a law might be as valid for the families of selected ancestry as Mendel's laws are for his peas and for other peas of probably similar ancestral history, but it would fail when applied to dark and light-eyed parents in general,—that is, to parents of any ancestry who happen to possess eyes of given colour."

The suggestion amounts to this: that because there are exceptions to dominance in peas; and because by some stupendous coincidence, or still more amazing incompetence, a bungler might have thought he found dominance of one eye-colour whereas really there was none*; therefore

* I have no doubt there is no universal dominance in eye-colour. Is it *quite* certain there is no dominance at all? I have searched the works of Galton and Pearson relating to this subject without finding a clear proof. If there is in them material for this decision

Professor Weldon is at liberty to suggest there is a fair chance that Mendel and all who have followed him have either been the victims of this preposterous coincidence not once, but again and again; or else persisted in the same egregious and perfectly gratuitous blunder. Professor Weldon is skilled in the Calculus of Chance: will he compute the probabilities in favour of his hypothesis?

Ancestry and purity of germ-cells.

To what extent ancestry is likely to elucidate dominance we have now seen. We will briefly consider how laws derived from ancestry stand in regard to segregation of characters among the gametes.

For Professor Weldon suggests that his view of ancestry will explain the facts not only in regard to dominance and its fluctuations but in regard to the purity of the germ-cells. He does not apply this suggestion in detail, for its error would be immediately exposed. In every strictly Mendelian case the *ancestry* of the pure extracted recessives or dominants, arising from the breeding of first crosses, is identical with that of the impure dominants [or impure recessives in cases where they exist]. Yet the posterity of each is wholly different. The pure extracted forms, in these simplest cases, are no more likely to produce the form with which they have been crossed than was their pure grandparent; while the impure forms break up again into both grand-parental forms.

Ancestry does not touch these facts in the least. They

I may perhaps be pardoned for failing to discover it, since the tabulations are not prepared with this point in view. Reference to the original records would soon clear up the point.

and others like them have been a stumbling-block to all naturalists. Of such paradoxical phenomena Mendel now gives us the complete and final account. Will Professor Weldon indicate how he proposes to regard them?

Let me here call the reader's particular attention to that section of Mendel's experiments to which Professor Weldon does not so much as allude. Not only did Mendel study the results of allowing his cross-breds (*DR*'s) to fertilise themselves, giving the memorable ratio

$$1\,DD : 2\,DR : 1\,RR,$$

but he fertilised those cross-breds (*DR*'s) both with the pure dominant (*D*) and with the pure recessive (*R*) varieties reciprocally, obtaining in the former case the ratio

$$1\,DD : 1\,DR$$

and in the latter the ratio

$$1\,DR : 1\,RR.$$

The *DD* group and the *RR* group thus produced giving on self-fertilisation pure *D* offspring and pure *R* offspring respectively, while the *DR* groups gave again

$$1\,DD : 2\,DR : 1\,RR.$$

How does Professor Weldon propose to deal with these results, and by what reasoning can he suggest that considerations of ancestry are to be applied to them? If I may venture to suggest what was in Mendel's mind when he applied this further test to his principles it was perhaps some such considerations as the following. Knowing that the cross-breds on self-fertilisation give

$$1\,DD : 2\,DR : 1\,RR$$

three explanations are possible:

(*a*) These cross-breds may produce pure *D* germs of both sexes and pure *R* germs of both sexes on an average in equal numbers.

(*b*) *Either* the female, *or* the male, gametes may be *alone* differentiated according to the allelomorphs, into pure *D*'s, pure *R*'s, and crosses *DR* or *RD*, the gametes of the other sex being homogeneous and neutral in regard to those allelomorphs.

(*c*) There may be some neutralisation or cancelling between characters in *fertilisation* occurring in such a way that the well-known ratios resulted. The absence of and inability to transmit the *D* character in the *RR*'s, for instance, might have been due not to the original purity of the germs constituting them, but to some condition incidental to or connected with fertilisation.

It is clear that Mendel realized (*b*) as a possibility, for he says *DR* was fertilised with the pure forms to test the composition of its egg-cells, but the reciprocal crosses were made to test the composition of the pollen of the hybrids. Readers familiar with the literature will know that both Gärtner and Wichura had in many instances shown that the offspring of crosses in the form $(a \times b)$ ♀ × c ♂ were less variable than those of crosses in the form a ♀ × $(b \times c)$ ♂, &c. This important fact in many cases is observed, and points to differentiation of characters occurring frequently among the male gametes when it does not occur or is much less marked among the maternal gametes. Mendel of course knew this, and proceeded to test for such a possibility, finding by the result that differentiation was the same in the gametes of both sexes*.

* See Wichura (46), pp. 55–6.

Of hypotheses (*b*) and (*c*) the results of recrossing with the two pure forms dispose; and we can suggest no hypothesis but (*a*) which gives an acceptable account of the facts.

It is the purity of the "extracted" recessives and the "extracted" dominants—primarily the former, as being easier to recognize—that constitutes the real proof of the validity of Mendel's principle.

Using this principle we reach immediately results of the most far-reaching character. These theoretical deductions cannot be further treated here—but of the practical use of the principle a word may be said. Where-ever there is marked dominance of one character the breeder can at once get an indication of the amount of trouble he will have in getting his cross-bred true to either dominant or recessive character. He can only thus forecast the future of the race in regard to each such pair of characters taken severally, but this is an immeasurable advance on anything we knew before. More than this, it is certain that in some cases he will be able to detect the "mule" or heterozygous forms by the statistical frequency of their occurrence or by their structure, especially when dominance is absent, and sometimes even in cases where there is distinct dominance. With peas, the practical seedsman cares, as it happens, little or nothing for those simple characters of seed-structure, &c. that Mendel dealt with. He is concerned with size, fertility, flavour, and numerous similar characters. It is to these that Laxton (invoked by Professor Weldon) primarily refers, when he speaks of the elaborate selections which are needed to fix his novelties.

We may now point tentatively to the way in which some even of these complex cases may be elucidated by an extension of Mendel's principle, though we cannot forget that there are other undetected factors at work.

The value of the appeal to Ancestry.

But it may be said that Professor Weldon's appeal to ancestry calls for more specific treatment. When he suggests ancestry as "one great reason" for the different properties displayed by different races or individuals, and as providing an account of other special phenomena of heredity, he is perhaps not to be taken to mean any definite ancestry, known or hypothetical. He may, in fact, be using the term "ancestry" merely as a brief equivalent signifying the previous history of the race or individual in question. But if such a plea be put forward, the real utility and value of the appeal to ancestry is even less evident than before.

Ancestry, as used in the method of Galton and Pearson, means a definite thing. The whole merit of that method lies in the fact that by it a definite accord could be proved to exist between the observed characters and behaviour of specified descendants and the ascertained composition of their pedigree. Professor Weldon in now attributing the observed peculiarities of *Telephone* &c. to conjectural peculiarities of pedigree—if this be his meaning—renounces all that had positive value in the reference to ancestry. His is simply an appeal to ignorance. The introduction of the word "ancestry" in this sense contributes nothing. The suggestion that ancestry might explain peculiarities means no more than "we do not know how peculiarities are to be explained." So Professor Weldon's phrase "peas of probably similar ancestral history*" means "peas probably

* See above, p. 192.

similar"; when he speaks of Mendel having obtained his results with "a few pairs of plants of known ancestry*," he means "a few pairs of known plants" and no more; when he writes that "the law of segregation, like the law of dominance appears to hold only for races of particular ancestry†," the statement loses nothing if we write simply "for particular races." We all know—the Mendelian, best of all—that particular races and particular individuals may, even though indistinguishable by any other test, exhibit peculiarities in heredity.

But though on analysis those introductions of the word "ancestry" are found to add nothing, yet we can feel that as used by Professor Weldon they are intended to mean a great deal. Though the appeal may be confessedly to ignorance, the suggestion is implied that if we did know the pedigrees of these various forms we should then have some real light on their present structure or their present behaviour in breeding. Unfortunately there is not the smallest ground for even this hope.

As Professor Weldon himself tells us‡, conclusions from pedigree must be based on the conditions of the several ancestors; and even more categorically (p. 244), "*The degree to which a parental character affects offspring depends not only upon its development in the individual parent, but on its degree of development in the ancestors of that parent.*" [My italics.] Having rehearsed this profession of an older faith Professor Weldon proceeds to stultify it in his very next paragraph. For there he once again reminds us that *Telephone*, the mongrel pea of recent origin, which does not breed true to seed characters, has yet manifested the peculiar power of stamping the recessive characters on its cross-bred

* See above, p. 187. † See above, p. 184.
‡ See above, p. 186.

offspring, though pure and stable varieties that have exhibited the same characters in a high degree for generations have *not* that power. As we now know, the presence or absence of a character in a progenitor *may* be no indication whatever as to the probable presence of the character in the offspring; for the characters of the latter depend on gametic and not on zygotic differentiation.

The problem is of a different order of complexity from that which Professor Weldon suggests, and facts like these justify the affirmation that if we could at this moment bring together the whole series of individuals forming the pedigree of *Telephone*, or of any other plant or animal known to be aberrant as regards heredity, we should have no more knowledge of the nature of these aberrations; no more prescience of the moment at which they would begin, or of their probable modes of manifestation; no more criterion in fact as to the behaviour such an individual would exhibit in crossing*, or solid ground from which to forecast its posterity, than we have already. We should learn then—what we know already—that at some particular point of time its peculiar constitution was created, and that its peculiar properties then manifested themselves: how or why this came about, we should no more comprehend with the full ancestral series before us, than we can in ignorance of the ancestry. Some cross-breds follow Mendelian segregation; others do not. In some, palpable dominance appears; in others it is absent.

If there were no ancestry, there would be no posterity. But to answer the question *why* certain of the posterity depart from the rule which others follow, we must know, not the ancestry, but how it came about *either* that at a

* Beyond an indication as to the homogeneity or "purity" of its gametes at a given time.

certain moment a certain gamete divided from its fellows in a special and unwonted fashion; *or*, though the words are in part tautological, the reason why the union of two particular gametes in fertilisation took place in such a way that gametes having new specific properties resulted*. No one yet knows how to use the facts of ancestry for the elucidation of these questions, or how to get from them a truth more precise than that contained in the statement that a diversity of specific consequences (in heredity) may follow an apparently single specific disturbance. Rarely even can we see so much. The appeal to ancestry, as introduced by Professor Weldon, masks the difficulty he dare not face.

In other words, it is the *cause of variation* we are here seeking. To attack that problem no one has yet shown the way. Knowledge of a different order is wanted for that task; and a compilation of ancestry, valuable as the exercise may be, does not provide that particular kind of knowledge.

Of course when once we have discovered by experiment that—say, *Telephone*—manifests a peculiar behaviour in heredity, we can perhaps make certain forecasts regarding it with fair correctness; but that any given race or individual will behave in such a way, is a fact not deducible from its ancestry, for the simple reason that organisms of identical ancestry may behave in wholly distinct, though often definite, ways.

It is from this hitherto hopeless paradox that Mendel has begun at last to deliver us. The appeal to ancestry is a substitution of darkness for light.

* May there be a connection between the extraordinary fertility and success of the *Telephone* group of peas, and the peculiar frequency of a blended or mosaic condition of their allelomorphs? The conjecture may be wild, but it is not impossible that the two phenomena may be interdependent.

VII. The question of absolute purity of germ-cells.

But let us go back to the cases of defective "purity" and consider how the laws of ancestry stand in regard to them. It appears from the facts almost certain that purity may sometimes be wanting in a character which elsewhere usually manifests it.

Here we approach a question of greater theoretical consequence to the right apprehension of the part borne by Mendelian principles in the physiology of heredity. We have to consider the question whether the purity of the gametes in respect of one or other antagonistic character is or is likely to be in case of *any* given character a *universal* truth? The answer is unquestionably—No—but for reasons in which "ancestry" plays no part*.

Hoping to interest English men of science in the Mendelian discoveries I offered in November 1900 a paper on this subject to "Nature." The article was of some length and exceeded the space that the Editor could grant without delay. I did not see my way to reduce it without injury to clearness, and consequently it was returned to me. At the time our own experiments were not ready for publication and it seemed that all I had to say would probably be common knowledge in the next few weeks, so no further attempt at publication was made.

In that article I discussed this particular question of the absolute purity of the germ-cells, showing how, on the analogy of other bud-variations, it is almost certain that the germ-cells, even in respect to characters normally Mendelian, may on occasion present the same mixture of characters, whether apparently blended or mosaic, which

* This discussion leaves "false hybridism" for separate consideration.

we know so well elsewhere. Such a fact would in nowise diminish the importance of Mendel's discovery. The fact that mosaic peach-nectarines occur is no refutation of the fact that the *total* variation is common. Just as there may be trees with several such mosaic fruits, so there may be units, whether varieties, individual plants, flowers or gonads, or other structural units, bearing mosaic egg-cells or pollen grains. Nothing is more likely or more in accordance with analogy than that by selecting an individual producing germs of blended or mosaic character, a race could be established continuing to produce such germs. Persistence of such blends or mosaics in *asexual* reproduction is well-known to horticulturists; for example "bizarre" carnations, oranges streaked with "blood"-orange character, and many more. In the famous paper of Naudin, who came nearer to the discovery of the Mendelian principle than any other observer, a paper quoted by Professor Weldon, other examples are given. These forms, once obtained, can be multiplied *by division*; and there is no reason why a zygote formed by the union of mosaic or blended germs, once arisen, should not in the cell-divisions by which its gametes are formed, continue to divide in a similar manner and produce germs like those which united to form that zygote. The irregularity, once begun, may continue for an indefinite number of divisions.

I am quite willing to suppose, with Professor Weldon (p. 248), that the pea *Stratagem* may, as he suggests, be such a case. I am even willing to accept provisionally as probable that when two gametes, themselves of mosaic or blended character, meet together in fertilisation, they are more likely to produce gametes of mosaic or blended character than of simply discontinuous character. Among Messrs Sutton's Primulas there are at least two striking cases of "flaked" or "bizarre" unions of bright colours and white which reproduce themselves by seed with fair constancy, though Mendelian purity in respect of these colours is elsewhere common in the varieties (I suspect mosaics of "false hybridism" among allelomorphs in some of these cases). Similarly Galton has shown that though children having one light-eyed and one dark-eyed parent generally have eyes either light or dark, the comparatively rare medium eye-coloured persons when they mate together frequently produce children with medium eye-colour.

In this connection it may be worth while to allude to a point of some practical consequence. We know that when pure dominant—say yellow—is crossed with pure recessive—say green—the dominance of yellow is seen; and we have every reason to believe this rule generally (*not* universally) true for pure varieties of peas. But we notice that in the case of a form like the pea, depending on human selection for its existence, it might be possible in a few years for the races with pure seed characters to be practically supplanted by the "mosaicized" races like the *Telephone* group, if the market found in these latter some specially serviceable quality. In the maincrop peas I suspect this very process is taking place*. After such a

* Another practical point of the same nature arises from the great variability which these peas manifest in plant- as well as seed-characters. Mr Hurst of Burbage tells me that in *e.g. William the First*, a pea very variable in seed-characters also, tall plants may be so common that they have to be rogued out even when the variety is grown for the vegetable market, and that the same is true of several such varieties. It seems by no means improbable that it is by such roguing that the unstable mosaic or blend-form is preserved. In a thoroughly stable variety such as *Ne Plus Ultra* roguing is hardly necessary even for the seed-market.

Mr N. N. Sherwood in his useful account of the origin and races revolution it might be possible for a future experimenter to conclude that *Pisum sativum* was by nature a "mosaicized" species in these respects, though the mosaic character may have arisen once in a seed or two as an exceptional phenomenon. When the same reasoning is extended to wild forms depending on other agencies for selection, some interesting conclusions may be reached.

But in Mendelian cases we are concerned primarily not with the product of gametes of blended character, but with the consequences of the union of gametes already discontinuously dissimilar. The existence of pure Mendelian gametes for given characters is perfectly compatible with the existence of blended or mosaic gametes for similar characters elsewhere, but this principle enables us to form a comprehensive and fruitful conception of the relation of the two phenomena to each other. As I also pointed out, through the imperfection of our method which does not yet permit us to *see* the differentiation among the gametes though we know it exists, we cannot yet as a rule obtain certain proof of the impurity of the gametes (except perhaps in the case of mosaics) as distinct from evidence of imperfect dominance. If however the case be one of a "mule" form, distinct from either parent, and not merely of dominance, there is no *a priori* reason why even this may not be possible; for we should be able to

of peas (*Jour. R. Hort. Soc.* XXII. 1899, p. 254) alludes to the great instability of this class of pea. To Laxton, he says, "we are indebted for a peculiar type of Pea, a round seed with a very slight indent, the first of this class sent out being *William the First*, the object being to get a very early blue-seeded indented Pea of the same earliness as the Sangster type with a blue seed, or in other words with a Wrinkled Pea flavour. This type of Pea is most difficult to keep true on account of the slight taint of the Wrinkled Pea in the breed, which causes it to run back to the Round variety."

distinguish the results of breeding first crosses together into *four* classes: two pure forms, one or more blend or mosaic forms, and "mule" forms. Such a study could as yet only be attempted in simplest cases: for where we are concerned with a compound allelomorph capable of resolution, the combinations of the integral components become so numerous as to make this finer classification practically inapplicable.

But in many cases—perhaps a majority—though by Mendel's statistical method we can perceive the fluctuations in the numbers of the several products of fertilisation, we shall not know whether abnormalities in the distribution of those products are due to a decline in dominance, or to actual impurity of the gametes. We shall have further to consider, as affecting the arithmetical results, the possibility of departure from the rule that each kind of gamete is produced in equal numbers*; also that there may be the familiar difficulties in regard to possible selection and assortative matings among the gametes.

I have now shown how the mosaic and blend-forms are to be regarded in the light of the Mendelian principle. What has Professor Weldon to say in reference to them? His suggestion is definite enough—that a study of ancestry will explain the facts: *how*, we are not told.

In speaking of the need of study of the characters of the *race* he is much nearer the mark, but when he adds "that is their ancestry," he goes wide again. When *Telephone* does not truly divide the antagonistic characters among its germ-cells this fact is in nowise simply traceable to its having originated in a cross—a history it shares with almost all the peas in the market—but to its own peculiar

* In dealing with cases of decomposition or resolution of compound characters this consideration is of highest importance.

nature. In such a case imperfect dominance need not surprise us.

What we need in all these phenomena is a knowledge of the properties of each race, or variety, as we call it in peas. We must, as I have often pleaded, study the properties of each form no otherwise than the chemist does the properties of his substances, and thus only can we hope to work our way through these phenomena. *Ancestry* holds no key to these facts; for the same ancestry is common to own brothers and sisters endowed with dissimilar properties and producing dissimilar posterity. To the knowledge of the properties of each form and the laws which it obeys there are no short cuts. We have no periodic law to guide us. Each case must as yet be separately worked out.

We can scarcely avoid mention of a further category of phenomena that are certain to be adduced in opposition to the general truth of the purity of the extracted forms. It is a fact well known to breeders that a highly-bred stock may, unless selections be continued, "degenerate." This has often been insisted on in regard to peas. I have been told of specific cases by Messrs Sutton and Sons, instances which could be multiplied. Surely, will reply the supporters of the theory of Ancestry, this is simply impurity in the extracted stocks manifesting itself at last. Such a conclusion by no means follows, and the proof that it is inapplicable is obtained from the fact that the "degeneration," or variation as we should rather call it, need not lead to the production of any proximate ancestor of the selected stock at all, but immediately to a new form, or to one much more remote—in the case of some high class peas, *e.g.*, to the form which Mr Sutton describes as "vetch-like," with short pods, and a very few small round seeds, two or three in a pod. Such plants are recognized by their appearance and are rigorously hoed out every year before seeding.

To appreciate the meaning of these facts we must go back to what was said above on the nature of compound characters. We can perceive that, as Mendel showed, the integral characters of the varieties can be dissociated and re-combined in any combination. More than that; certain integral characters can be resolved into further integral components, by *analytical* variations. What is taking place in this process of resolution we cannot surmise, but we may liken the consequences of that process to various phenomena of analysis seen elsewhere. To continue the metaphor we may speak of return to the vetch-like type as a *synthetical* variation: well remembering that we know nothing of any *substance* being subtracted in the former case or added in the latter, and that the phenomenon is more likely to be primarily one of alteration in arrangement than in substance.

A final proof that nothing is to be looked for from an appeal to ancestry is provided by the fact—of which the literature of variation contains numerous illustrations—that such newly synthesised forms, instead of themselves producing a large proportion of the high class variety which may have been their ancestor for a hundred generations, may produce almost nothing but individuals like themselves. A subject fraught with extraordinary interest will be the determination whether by crossing these newly synthesised forms with their parent, or another pure form, we may not succeed in reproducing a great part of the known series of components afresh. The pure parental form, produced, or extracted, by "analytical" breeding, would not in ordinary circumstances be capable of producing the other components from which it has been separated; but by crossing it with the "synthesised" variety it is not impossible that these components would again reappear. If this can be shown to be possible we shall have entirely new light on the nature of variation and stability.

Conclusion.

I trust what I have written has convinced the reader that we are, as was said in opening, at last beginning to move. Professor Weldon declares he has "no wish to belittle the importance of Mendel's achievement"; he desires "simply to call attention to a series of facts which seem to him to suggest fruitful lines of inquiry." In this purpose I venture to assist him, for I am disposed to think that unaided he is—to borrow Horace Walpole's phrase—about as likely to light a fire with a wet dish-clout as to kindle interest in Mendel's discoveries by his tempered appreciation. If I have helped a little in this cause my time has not been wasted.

In these pages I have only touched the edge of that new country which is stretching out before us, whence in ten years' time we shall look back on the present days of our captivity. Soon every science that deals with animals and plants will be teeming with discovery, made possible by Mendel's work. The breeder, whether of plants or of animals, no longer trudging in the old paths of tradition, will be second only to the chemist in resource and in foresight. Each conception of life in which heredity bears a part—and which of them is exempt?—must change before the coming rush of facts.

BIBLIOGRAPHY.

1. CORRENS, C. G. Mendel's Regel über das Verhalten der Nachkommenschaft der Rassenbastarde, *Ber. deut. bot. Ges.*, XVIII., 1900, p. 158.
2. —— Gregor Mendel's "Versuche über Pflanzen-Hybriden" und die Bestätigung ihrer Ergebnisse durch die neuesten Untersuchungen, *Bot. Ztg.*, 1900, p. 229.
3. —— Ueber Levkoyenbastarde zur Kenntniss der Grenzen der Mendel'schen Regeln, *Bot. Cblt.*, 1900, Vol. LXXXIV., p. 97.
4. —— Bastarde zwischen Maisrassen, mit besonderer Berücksichtigung der Xenien, *Bibliotheca Botanica*, Hft. 53, 1901.
5. CRAMPE. Kreuzungen zwischen Wanderratten verschiedener Farbe, *Landwirths. Jahrb.*, VI., 1877, p. 384.
6. —— Zucht-Versuche mit zahmen Wanderratten. 1. Resultate der Zucht in Verwandtschaft, *ibid.*, XII., 1883, p. 389. 2. Resultate der Kreuzung der zahmen Ratten mit wilden, *ibid.*, XIII., 1884, p. 699.
7. —— Die Gesetze der Vererbung der Farbe, *ibid.*, XIV., p. 539.
8. DARWIN, C. *Variation of Animals and Plants under Domestication*, ed. 2, I., pp. 348 and 428.
9. FISCHER, JOHANN VON. Die Säugethiere des St Petersburger Gouvernements, *Zool. Garten*, X., 1869, p. 336.
10. —— Iltis (*Mustela putorius*) und Frett (*Mustela furo*), *ibid.*, XIV., 1873, p. 108.

11. Fischer, Johann von. Beobachtungen über Kreuzungen verschiedener Farbenspielarten innerhalb einer Species, *ibid.*, xv., 1874, p. 361.
12. Focke, W. O. *Die Pflanzen-Mischlinge*, Bornträger, Berlin, 1881.
13. —— Ueber dichotype Gewächse. *Oesterr. bot. Ztschr.*, xviii., 1868, p. 139.
14. Galton, F. *Natural Inheritance*, Macmillan and Co., London, 1889.
15. —— The Average Contribution of each several Ancestor to the total Heritage of the Offspring, *Proc. Roy. Soc.*, lxi., 1897, p. 401.
16. Gärtner, C. F. von. *Versuche und Beobachtungen über die Bastarderzeugung im Pflanzenreich*, Stuttgart, 1849.
17. Gitay, E. Ueber den directen Einfluss des Pollens auf Frucht- und Samenbildung, *Pringsheim's JB. d. wiss. Bot.*, xxv., 1893, p. 489.
18. Godron, D. A. Des Hybrides Végétaux, etc. *Ann. Sci. Nat. Bot.*, Ser. 4, xix., 1863, p. 135, and a series of papers in *Mém. Acad. Stanislas*, Nancy, 1864, 1865, and especially 1872.
19. Guaita, G. von. Versuche mit Kreuzungen von verschiedenen Rassen der Hausmaus, *Ber. d. naturf. Ges. Freiburg*, x., 1898, p. 317.
20. —— Zweite Mittheilung, etc., *ibid.*, xi., 1900, p. 131.
21. Knight, T. A. An account of some experiments on the Fecundation of Vegetables, *Phil. Trans.*, 1799, Pt. ii., p. 195.
22. Küster, E. Die Mendel'schen Regeln, ihre ursprüngliche Fassung und ihre moderne Ergänzungen, *Biol. Cblt.*, xxii., 1902, p. 129.
23. Laxton, T. Observations on the variations effected by crossing in the colour and character of the seed of Peas, *Internat. Hort. Exhib. and Bot. Congr.*, Report, 1866, p. 156.
24. —— Notes on some Changes and Variations in the

Offspring of Cross-fertilized Peas, *Jour. Hort. Soc.*, N.S. III., 1872, p. 10.

25. LAXTON, T. Improvement amongst Peas, *Jour. Hort. Soc.*, 1890, XII., 1, p. 29.
26. MENDEL, GREGOR JOHANN. Versuche über Pflanzen-Hybriden, *Verh. naturf. Ver. in Brünn*, Band IV., 1865, *Abhandlungen*, p. 1; reprinted in *Flora*, 1901, and in Ostwald's *Klassiker d. exakten Wiss.* English translation in *Jour. R. Hort. Soc.*, 1901, XXVI.
27. —— Ueber einige aus künstlicher Befruchtung gewonnenen Hieracium-Bastarde, *ibid.*, VIII., 1869, *Abhandlungen*, p. 26.
28. MILLARDET. Note sur l'hybridation sans croisement, ou fausse hybridation, *Mém. Soc. Sci. Bordeaux*, Ser. 4, IV., 1894, p. 347.
29. C. NAUDIN. Nouvelles recherches sur l'Hybridité dans les Végétaux, *Nouv. Arch. Mus.*, I., 1865, p. 25.
30. —— *Ann. sci. nat.*, *Bot.*, Ser. 4, XIX., p. 180.
31. PEARSON, KARL. On the Law of Ancestral Heredity, *Proc. Roy. Soc.*, LXII., 1898, p. 386.
32. —— On the Law of Reversion, *ibid.*, LXVI., 1900, p. 140.
33. —— *The Grammar of Science*, second edition, London, A. and Charles Black, 1900.
34. —— Mathematical Contributions to the Theory of Evolution. VIII. On the Inheritance of Characters not capable of exact Measurement, *Phil. Trans. Roy. Soc.*, 1900, Vol. 195, p. 79.
35. RIMPAU. Kreuzungsprodukte landw. Kulturpflanzen, *Landw. Jahrb.*, XX., 1891.
36. TSCHERMAK, E. Ueber künstliche Kreuzung bei *Pisum sativum*, *Ztschrft. f. d. landwirths. Versuchswesen in Oesterr.*, 1900, III., p. 465.
37. —— Weitere Beiträge über Verschiedenwerthigkeit der Merkmale bei Kreuzung von Erbsen und Bohnen, *ibid.*, 1901, IV., 641; *abstract* in *Ber. deut. bot. Ges.*, 1901, XIX., p. 35.

38. TSCHERMAK, E. Ueber Züchtung neuer Getreiderassen mittelst künstlicher Kreuzung, *ibid.*, 1901, IV., p. 1029.
39. VILMORIN-ANDRIEUX AND Co. *Les Plantes Potagères*, 1st ed. 1883; 2nd ed. 1891.
40. VRIES, H. DE. Sur la loi de disjonction des hybrides, *Comptes Rendus*, 26 March, 1900.
41. —— Das Spaltungsgesetz der Bastarde, *Ber. deut. bot. Ges.*, 1900, XVIII., p. 83.
42. —— Ueber erbungleiche Kreuzungen, *ibid.*, p. 435.
43. —— Sur les unités des caractères spécifiques et leur application à l'étude des hybrides, *Rev. Gén. de Bot.*, 1900, XII., p. 257. See also by the same author, *Intracellulare Pangenesis*, Jena, 1889, in which the conception of unit-characters is clearly set forth.
44. —— *Die Mutationstheorie*, Vol. I., Leipzig, 1901.
45. WELDON, W. F. R. Mendel's Laws of Alternative Inheritance in Peas, *Biometrika*, I., Pt. ii., 1902, p. 228.
46. WICHURA, MAX. Die Bastardbefruchtung im Pflanzenreich, erläutert an den Bastarden der Weiden, Breslau, 1865.

Received as this sheet goes to press:—

CORRENS, C. Die Ergebnisse der neuesten Bastardforschungen für die Vererbungslehre, *Ber. deut. bot. Ges.*, XIX., Generalversammlungs-Heft 1.

—— Ueber den Modus und den Zeitpunkt der Spaltung der Anlagen bei den Bastarden vom Erbsen-Typus, *Bot. Ztg.*, 1902, p. 65.

CAMBRIDGE: PRINTED BY J. AND C. F. CLAY, AT THE UNIVERSITY PRESS.

Printed in the United States
By Bookmasters